T5-AFN-111

ENVIRONMENTAL GEOCHEMISTRY AND HEALTH

The GeoJournal Library

Series Editor: WOLF TIETZE

Environmental Geochemistry and Health

Report to the Royal Society's British National Committee for Problems of the Environment

Edited by

S. H. U. Bowie, F.R.S.

Visiting Professor of Applied Geology,
University of Strathclyde, Glasgow

and

I. Thornton

Reader in Environmental Geochemistry,
Applied Geochemistry Research Group,
Imperial College, London

D. Reidel Publishing Company

A MEMBER OF THE KLUWER ACADEMIC PUBLISHERS GROUP

Dordrecht / Boston / Lancaster

Library of Congress Cataloging in Publication Data
Main entry under title:

CIP

Environmental geochemistry and health.

(GeoJournal library)
Includes bibliographies and index.
1. Environmental health—Congresses. 2. Trace elements—Environmental aspects—Congresses. 3. Trace elements—Physiological effect—Congresses. 4. Medical geography—Congresses. 5. Biogeochemistry—Congresses. I. Bowie, S. H. U. (Stanley Hay Umphray), 1917- II. Thornton, Iain. III. British National Committee for Problems of the Environment. IV. Series. [DNLM: 1. Environmental Exposure. 2. Trace Elements—adverse effects. 3. Trace Elements—deficiency. QU 130 E61]
RA 565.A2E575 1984 574.2'4 84-18253
ISBN 90-277-1879-2

Published by D. Reidel Publishing Company
P.O. Box 17, 3300 AA Dordrecht, Holland

Sold and distributed in the U.S.A. and Canada
by Kluwer Academic Publishers,
190 Old Derby Street, Hingham, MA 02043, U.S.A.

In all other countries, sold and distributed
by Kluwer Academic Publishers Group,
P.O. Box 322, 3300 AH Dordrecht, Holland

Printed in the Netherlands

TABLE OF CONTENTS

PREFACE

1. INTRODUCTION 1
   1.1. Origins and Remit 1
   1.2. Actions Taken 2
   1.3. Royal Society Working Party on Environmental Geochemistry and Health 3

2. PRINCIPLES OF ENVIRONMENTAL GEOCHEMISTRY 5
   2.1. Summary 5
   2.2. Introduction 6
   2.3. The Distribution of Elements in Rocks and some Geochemical Associations 7
   2.4. Redistribution of Chemical Elements by Weathering 13
   2.5. Chemical Elements in the Surface Environment and Factors Influencing Redistribution 15
   2.6. Uranium and Daughters 18
   2.7. Regional Geochemical Maps 19
   2.8. Regional Geochemistry of Britain 25
   2.9. Trace Elements in Soils 28
   2.10. Metal Pollution 32
   2.11. References 32

3. PLANT-SOIL PROCESSES 35
   3.1. Summary 35
   3.2. Introduction 37

3.3. Major Elements N, P, K, Ca, and Mg 37
3.4. Other Elements 39
3.5. The Availability of Elements: Plant Factors 40
3.6. The Availability of Elements in the Soil: Soil Factors 47
3.7. The Root-Soil System 51
3.8. Practical Crop Problems in Britain 51
3.9. The Availability and Need for Soil Information Relating to Geochemistry 55
3.10. References 56

4. GEOCHEMISTRY AND ANIMAL HEALTH 59
4.1. Summary 59
4.2. Introduction 61
4.3. Inorganic Element Deficiency Diseases 62
4.4. Inorganic Element Toxicity 65
4.5. Dietary Requirements 67
4.6. Factors Modifying the Response of Animals to their Geochemical Environment 67
4.7. Prediction of the Risks of Deficiency or Toxicity from Geochemical Survey Data 77
4.8. The Influence of Geochemical Anomalies upon Animal Health: An Appraisal of Field Evidence 78
4.9. Conclusions 89
4.10. References 90
4.11. Appendix: Relations between the Distribution of High-Mo Geochemical Anomalies and of Cu Deficiency in Cattle in Caithness, N.E. Sutherland 92

5. GEOCHEMISTRY AND HUMAN HEALTH 97

5.1. Summary 97

5.2. Introduction 99

5.3. Iodine and Fluorine 100

5.4. Cardiovascular Disease 102

5.5. Calcium, Magnesium, and Sodium 105

5.6. Cancers and Geochemistry 108

5.7. Multiple Sclerosis and Geochemistry 109

5.8. Anomalous Geochemical Areas 109

5.9. Trace Elements and Health 112

5.10. Blood Levels of Trace Elements 117

5.11. References 118

6. CONCLUSIONS 121

7. RECOMMENDATIONS 127

INDEX 131

PREFACE

One of the main outcomes of the eleven meetings of the Working Party was the recognition of the importance of interdisciplinary studies linking regional geochemistry with plant, animal and human health.

The effects of major element deficiencies or excesses on plant health are well known; this is not the case for trace elements. In fact, rapid and reliable analytical methods for determining trace element abundances have only recently become available, and it is to be expected that important new information on trace element levels will be forthcoming. This, however, is only part of the problem because other factors such as element speciation, uptake and transmission may be more significant than total concentration.

The pathways of elements from crops to animals are relatively well defined, but the aetiology of diseases attributable to elemental inadequacies or excesses is generally quite complex. Nevertheless, there is good evidence for diseases in livestock in the British Isles induced by deficiencies of Cu, Se and Co and Mo excess. On a world scale there is also convincing data on the effect of Na, P and I deficiencies and F excess on animal health. What is generally lacking, however, is adequate interaction between geochemists and biochemists, veterinary scientists and other concerned with animal health. Interpretation of geochemical data is complex as are connections between elemental abundances and the health of animals.

The links in the food chain from plant or animal to man are even more intricate since they involve food and water distribution and sometimes a large component attributable to atmospheric contamination as well as to social and economic factors. Even the well-known examples of the association between I deficiency and goitre, and F deficiency and dental caries are not fully explained at present. Neither is the general association of water hardness with cardiovascular disease. Cause and effect relationships require to be established, but these can scarcely be expected to be verified when so little is known about the geochemistry of the elements concerned or when speciation or interelement effects are not considered.

In all aspects of environmental geochemistry and health sub-clinical evidence of growth retardation or susceptibility to disease as well as the overt disease conditions themselves seem to be linked to elemental imbalances or deficiencies. These sub-clinical effects are though to be of particular economic significance to agricultural production. This, however, is an even more difficult field for investigation than a direct geochemistry-disease relationship.

There is little doubt that fresh geochemical data will throw new light on the relationship between geochemistry and health. What is not yet clear, however, is what geochemical data are necessary. This can only be arrived at by the closest possible collaboration between geochemists, soil scientists, biochemists, epidemiologists and other scientists. The Working Party has constituted the basis for such a combined approach. It is also essential that research be broadened to include information from other countries where relationships may be easier to establish because conditions are more extreme than in the British Isles. In this sphere the Royal Society could play a dominant role.

Readers may criticise the Working Party for the length of this Report, but members are convinced that presentation of the information in sections 2 to 5 in an unabridged form will provide valuable basic information for scientists in the various disciplines concerned and will go some way to explaining the complexities of the problems facing those who will be concerned with geochemistry and health in the future.

S. H. U. BOWIE, F.R.S.

1. INTRODUCTION

1.1. Origins and Remit

The British National Committee for Geology through its Geochemistry and Cosmochemistry Sub-Committee invited an *ad hoc* group, convened by Professor J. S. Webb, to consider the subject of geochemistry in relation to crop, animal and human health. Two meetings were held and a recommendation made to the Royal Society in March 1979 with the following terms of reference:

(i) to consider the existing role of studies in environmental geochemistry and health in national policies and, in particular:

(a) to investigate the extent of existing and likely future research in the United Kingdom on environmental geochemistry and health;

(b) to bring together researchers and users in the various disciplines in so far as their work relates to natural aspects of geochemistry both in the inorganic and organic field;

(c) to identify the needs and fields of research impinging on geochemistry in relation to health, agriculture and other disciplines in which important advances are likely to be made.

(ii) to advise on the possible role of the Royal Society in this field, not only nationally but in the international context.

1.2. Actions Taken

The Working Party first met on the 25th May, 1979 and thereafter on ten further occasions, the final meeting being on the 3rd December, 1981.

One of the first activities was to commence the compilation of a Register of Studies linking Health and the Geochemical Environment in the United Kingdom. This Register is now complete and is being issued separately.

Another early move was to make contact with existing organisations abroad concerned with environmental geochemistry and health. These included the Society for Environmental Geochemistry and Health in the U.S.A.; the U.S. Geological Survey, Regional Geochemistry Division; the U.S. Academy of Sciences Sub-Committee on Geochemistry and Health; National Research Council of Canada; Department of Animal Science, University of Western Australia; the Geochemistry Laboratory, Institute of Geochemistry, Academia Sinica, China; and the Norwegian Academy of Science and Letters.

In January 1981 a Discussion Meeting was organised to obtain the views of researchers in the United Kingdom concerned with geochemistry and health. Professor A. L. Page of the U.S. Academy of Sciences was also invited to give a review of developments in the U.S.A.

1.3. Royal Society Working Party On Environmental Geochemistry And Health

Chairman

Professor S. H. U. Bowie, F.R.S., Geological Consultant, Tanyard Farm, Clapton, Crewkerne, Somerset.

Vice-Chairman

Professor G. Eglinton, F.R.S., Professor of Organic Geochemistry, University of Bristol.

Secretary

Dr. I. Thornton, Reader in Environmental Geochemistry, Imperial College of Science and Technology.

Members

Dr. B. E. Davies, Senior Lecturer in Geography, University College of Wales, Aberystwyth.

Mr. W. Dermott, Head of the Agricultural Science Service, Agricultural Development and Advisory Service, MAFF.

Dr. F. A. Fairweather, Director of DHSS Toxicology Laboratories, replaced by Dr. M. F. Cuthbert, Principal Medical Officer, and later by Dr. G. K. Matthew, Principal Medical Officer, Toxicity and Environmental Pollution, DHSS.

Dr. N. J. King, Central Directorate on Environmental Pollution, Department of the Enviroment, replaced by Dr. D. L. Simms.

Dr. C. F. Mills, Head of Nutritional Biochemistry Department, Rowett Research Institute, Aberdeen.

Dr. Jane Plant, Principal Scientific Officer, Institute of Geological Sciences.

Professor A. G. Shaper, Professor of Clinical Epidemiology,

Royal Free Hospital School of Medicine.

Dr. P.B. Tinker, Head of Soils and Plant Nutrition Department, Rothamsted Experimental Station.

Professor J. S. Webb, Emeritus Professor of Applied Geochemistry, Imperial College of Science and Technology.

2. PRINCIPLES OF ENVIRONMENTAL GEOCHEMISTRY

2.1. Summary

Environmental geochemistry is concerned with the sources, distribution and interactions of chemical elements in the rock-soil-water-air-plant-animal-human systems. The primary source of elements are igneous rocks of which silicates and aluminosilicates are the dominant compounds. Trace elements commonly occur in accessory minerals but may be incorporated into the crystal lattices of silicates according to their valence and atomic radii. Sedimentary rocks are redeposited as a result of erosion and chemical alteration of pre-existing rocks. They comprise coarse and fine grained sediments, secondary phases such as clay minerals, and Fe and Mn oxides, and precipitates such as limestone.

In general, coarse grained sedimentary rocks such as sandstones contain low levels of trace elements: fine grained sediments such as shales, and particularly organic-rich black shales, contain large amounts. The chemical composition of metamorphic rocks usually reflects that of their sedimentary or igneous precursors. The redistribution of trace elements during metamorphism is poorly understood.

Elements are redistributed to varying degrees into the surface environment as a result of physical and chemical weathering of rock into soil. Chemical weathering is initiated mainly by interactions between rainwater and bedrock. The ease of weathering of rocks depends on their component minerals which may be resistant or soluble, and on the nature of the water with which they are in contact, which may range from high Ca or Mg 'hard' water to 'soft' water. Elements in resistate minerals are not chemically available, and with the exception of fibrous species, are unlikely to be of significance to health.

Elements in solution are mainly present as ions, ionic complexes and neutral molecules or organic and inorganic compounds. Whether elements are present as solid or liquid phases depends largely on the pH (H^+ ion activity) which affects processes such as ion exchange and complexing and on the Eh (redox potential) of a solution which measures its oxidising or reducing capacity. The behavour of elements is not only controlled by mineral-solution equilibrium but also by co-precipitation, surface effects such as adsorption and ion exchange, and interactions with organic phases. Both organic and inorganic compounds may occur as colloids in waters. These are important in metal transport as their surface charge interacts with dissolved ions. Colloids and chelating agents also form complexes with elements. Organic complexes, comprising in natural systems mainly fulvic and humic acids are of particular importance as they are able to contain considerable amounts of metal ions.

Uranium and its daugher products are both radiologically- and chemically-toxic elements. They are primarily concentrated in highly differentiated granitic and alkaline rocks. The parent element is very mobile as the uranyl ion and is readily concentrated by secondary processes in clastic sediments containing organic matter and iron sulphides. It is also closely associated with phosphatic sediments and lignites. The principal health hazard of the U series arises from the inhalation of radon which decays to the solid daughers ^{218}Po, ^{214}Po and ^{210}Po. These are carcinogenic mainly as a result of the damage done by their alpha-emitting solid decay products to the bronchial epithelium.

There is little systematic information in Britian on the geographic distribution of elements in soils, pasture herbage, food crops and water supplies and existing information is inadequate for regional studies on the health of livestock and human populations. The need for systematic data on trace element levels over Britain has been met, in part, by geochemical reconnaissance surveys carried out by Imperial College and the Institute of Geological Sciences.* These surveys are based mainly on the analysis of stream sediment samples by modern instrumental techniques for up to 35 chemical elements, and the use of automated methods to prepare maps showing the distribution of elements. Geochemical atlases have been published for Northern Ireland, England and Wales at 1 : 2M scale and parts of Scotland at 1 : 250,000 scale. Geochemical data for these and other areas are available in computer-readable form. These data may be interrelated with epidemiological data to explore possible relationships between geochemistry and health.

Geochemically, Britain can be divided into three physiographic/ geological regions:

(i) Precambrian crystalline basement rocks of Northern Scotland;
(ii) Palaeozoic shale/greywacke belts with granite intrusions of Southern Scotland, England and Wales;
(iii) Devonian-Tertiary sedimentary cover succession of England and South Wales.

The main sources of trace elements in British soils are weathered bedrock or overburden transported by wind, water or glaciation. Soils developed from these materials tend to reflect their chemical composition, though the influence of parent material is modified to varying degrees by soil forming processes, which may lead to the mobilisation and redistribution of elements. These processes inlude gleying, leaching, surface organic matter accumulation and podzolisation. Soils developed from acid igneous rocks and coarse grained sedimentary rocks usually contain smaller amounts of nutritionally essential trace elements than those derived from basic igneous and fine-grained sediments. Soils in mineralised areas frequently contain large amounts of potentially toxic metals which have often been dispersed beyond the original deposits in the course of mining and smelting activities. These and other pollutant sources of chemical elements in the soil are related to natural geochemical inputs and must be taken into account in studies of geochemistry and health.

2.2. Introduction

The distribution of elements in the Earth's crust is not

* Note: The Institute of Geological Sciences was renamed the British Geological Survey in January 1984.

random but is controlled by physico-chemical processes which are becoming better understood as a result of progress in geochemistry. Geochemistry is concerned with understanding the distribution of elements and their isotopes in the atmosphere, hydrosphere, crust, mantle and core of the Earth, but to man, the surface environment is of greatest importance. Environmental geochemistry has developed rapidly over the past decade mainly as a result of increased resource development and associated pollution. Much information on the processes controlling the distribution of elements in rocks and their dispersion and concentration in soil and water during weathering has been obtained, and systematic data on the levels of elements in the surface environment have also been accumulated by regional geochemical mapping. This information provides a basis for interdisciplinary studies in environmental geochemistry and health.

Environmental geochemistry is concerned with complex interactions in the system rock-soil-water-air-life which show marked variation from place to place. Here, three groups of elements are considered. Firstly, the major elements P, Ca, Mg, Na, K and Fe which are important not only to the health of crops, animals and man, but are also important controls on the primary distribution, and secondary dispersion of trace elements; secondly, the trace elements which are essential to animal life and which are the first row transition elements - Mn, (Fe), Ni, Cu, V, Co, Cr - together with Mo, Sn, Se, I, and F; and thirdly, such elements as Pb, Cd, Hg, and As which may have adverse physiological effects at relatively low levels.

2.3. The Distribution Of Elements In Rocks And Some Geochemical Associations

2.3.1. Igneous Rocks

The eight elements, O, Si, Al, Fe, Ca, Na, K, Mg in order of abundance, make up 99% of the Earth's crust. These elements form low density minerals of which silicates (and alumino-silicates) are dominant. Silicates formed at the highest temperatures and pressures are generally compounds of Fe, Mg, and Ca and occur in basic rocks, while those formed at the lowest temperatures and pressures are mostly compounds of Na, K or 'pure' silica (quartz) forming acid rocks such as granites. Incorporation of trace elements into the crystal lattices of silicates is controlled largely by the similarity of their valency and ionic radii to those of the major rock-forming elements. Hence the first row transition elements are mostly incorporated into Mg and Fe minerals in ultrabasic (Cr, Ni) or basic (Co, V) rocks. Pb^{2+} substitutes mostly for K^{+} while Mo, F, U, Sn, and W tend to be incompatible and are enriched in highly evolved K granites, in accessory minerals or micas. The averages abundances of trace elements in the most common types of igneous rocks are shown on Table I, and the most common igneous rock forming minerals are listed in Table Ia.

2.3.2. Sedimentary and Metamorphic Rocks

Igneous rocks, which are generated at high pressures and temperatures, are not in equilibrium with conditions at the Earth's surface and so are eroded and chemically altered. The weathering products are transported and redeposited as sediments variably composed of fragments of the parent material, and/or secondary phases such as clay minerals, and iron and manganese oxides. In addition, biochemical and chemical mineral precipitates may be formed, such as limestone, dolomite, phosphate and coal. The major element

TABLE I

Average abundance of some minor and trace elements in the Earth's crust, rocks and soil after Levinson 1974, and common igneous rock forming (all values in ppm).

Element	Earth's Crust	Ultra Basic	Basalt	Grano-diorite	Granite	Sand-stone	Shale	Lime-stone	Soil
Ag	0.07	0.06	0.1	0.07	0.04	-	0.05	1	0.1
As	1.8	1	2	2	1.5	1	15	2.5	1.50
Au	0.004	0.005	0.004	0.004	0.004	-	0.004	0.005	-
B	10	5	5	20	15	35	100	10	2-10
Ba	425	2	250	500	600	-	700	100	100-3000
Be	2.8	-	0.5	2	5	-	3	1	6
Bi	0.17	0.02	0.15	-	0.1	-	0.18	-	-
Br	2.5	1	3.6	-	2.9	1	4	6.2	-
Cd	0.2	-	0.2	0.2	0.2	-	0.2	0.1	1
Cl	130	85	60	-	165	10	180	150	-
Co	25	150	50	10	1	0.3	20	4	1-40
Cr	100	2000	200	20	4	35	100	10	5-1000
Cs	3	-	1	2	5	-	5	-	6
Cu	55	10	100	30	10	-	50	15	2-100

Table I (continued)

F	625	100	400	-	735	270	740	330	-
Ga	15	1	12	18	18	12	20	0.06	15
Ge	1.5	1	1.5	1	1.5	0.8	1.5	0.1	1
Hg	0.08	-	0.08	0.08	0.08	0.03	0.5	0.05	0.03
I	0.5	0.5	0.5	-	0.5	1.7	2.2	1.2	-
Li	20	-	10	25	30	15	60	20	5-200
Mn	950	1300	2200	1200	500	-	850	1100	850
Mo	1.5	0.3	1	1	2	0.2	3	1	2
Ni	75	2000	150	20	0.5	2	70	12	5-500
Pb	12.5	0.1	5	15	20	7	20	8	2-200
Rb	90	-	30	120	150	60	140	5	20-500
Sb	0.2	0.1	0.2	0.2	0.2	-	1	-	5
Se	0.05	-	0.05	-	0.05	0.05	0.6	0.08	0.2
Sn	2	0.5	1	2	3	-	4	4	10
Sr	375	1	465	450	285	20	300	500	50-1000
Te	0.001	0.001	0.001	0.001	0.001	-	0.01	-	-
Th	10	0.003	2.2	10	17	1.7	12	2	13
Ti	5700	3000	9000	8000	2300	1500	4600	400	5000
Tl	0.45	0.05	0.1	0.5	0.75	0.82	0.3	-	0.1
U	2.7	0.001	0.6	3	4.8	0.45	4	2	1
V	135	50	250	100	20	20	130	15	20-500
W	1.5	0.5	1	2	2	1.6	2	0.5	-
Zn	70	50	100	60	40	16	100	25	20

TABLE Ia

Oxides	SiO_2	quartz
	Fe_3O_4	magnetite
Feldspars	$NaAlSi_3O_8$	albite
	$KAlSi_3O_8$	orthoclase
	$CaAl_2Si_2O_8$	anorthite
Pyroxenes	$CaMgSi_2O_6$	diopside
	$MgSiO_3$	enstatite
Olivines	Mg_2SiO_4	forsterite
	Fe_2SiO_4	fayalite
Micas	$KAl_2AlSi_3O_{10}(OH)_2$	muscovite
	$KMg_3AlSi_3O_{10}(OH)_2$	phlogopite
Amphiboles	$Mg_7Si_8O_{22}(OH)_2$	anthophyllite
	$Ca_2Mg_5Si_8O_{22}(OH)_2$	actinolite

chemistry of the main types of sediment is shown in Table II after Fyffe (1974) and the trace element concentrations of some of the main groups of sediment are shown in Table I. The relatively low levels of trace elements over arenaceous sediments such as unmineralised sandstones and the correspondingly high levels over shales - particularly those associated with organic detritus (black shales) - are of particular significance in environmental geochemistry, while the calcium carbonate content of sediments is important in determining pH in the surface weathering regime.

Although such information provides general guidance on the levels of trace elements in different sedimentary

TABLE II

Major element geochemistry of sedimentary rocks after Fyffe 1974.

Rock type	Mineralogy	Chemistry
Sandstone	quartz, feldspars	dominated by SiO_2 (+K, Na, Ca, Al)
Shale	clay minerals, chlorites, carbonates	Al_2O_3-SiO_2-H_2O
Limestone	calcite, dolomite	$CaCO_3$, $MgCO_3$
Chert	quartz, haematite	SiO_2 (+ minor Fe_2O_3, MnO_2)
Phosphorite	apatite	calcium phosphate
Soil	complex clay minerals, quartz, organic materials	Al_2O_3-SiO_2-H_2O
Laterite	bauxite, haematite	Al_2O_3-Fe_2O_3-SiO_2
Evaporite	halite, gypsum	NaCl, $CaSO_4$, $CaCO_3$

lithologies, considerable variation may occur as a result of changes in the chemistry of the source region; physical and chemical conditions during weathering, transport and deposition; diagenesis; and in the case of sandstones, the movement of groundwater which may produce such high concentrations of U, V, Mo, and Se that ore deposits are formed. Reheating (metamorphism) of volcanic and sedimentary rocks occurs as a result of burial in tectonically active zones of the Earth's crust sometimes to depths of 30 km; temperatures may exceed 600 °C with pressures greater than 10,000 atmospheres. Most elements are thought not to be redistributed over large distances during metamorphism with the exception of H_2O, Cs, K, U, Th, and Rb and possibly B, which are depleted at the highest temperatures and pressures and under conditions of high partial pressure of CO_2. Thus, in general, the chemical composition of metamorphic rocks reflects that of their sedimentary or igneous precursors. The behaviour of trace elements and constraints on their redistribution during metamorphism are only poorly understood, however, and the average abundance of chemical elements in metamorphic assemblages is not well-documented.

2.4. Redistribution of Chemical Elements by Weathering

The redistribution of elements from bedrock into the surface environment occurs as a result of physical and chemical weathering which transforms rock which is frequently non-porous and of low reactivity, to soil which is porous and chemically active. Physical weathering breaks the rock into smaller particles, thereby increasing the surface area which is exposed to air and water, which are the main agents of chemical weathering.

Chemical weathering is initiated by interactions between rainwater and bedrock. Rainwater contains low concentrations of dissolved solids (Table III), which are mainly derived from the evaporation of seawater together with small amounts of atmospheric gases. A comparison between the average composition of rainwater and river water (Table III) shows that the latter contains a higher concentration of total dissolved solids (TDS) and different proportions of major elements - changes which result from such chemical weathering processes as dissolution, oxidation, hydrolysis and acid hydrolysis.

2.4.1. Influence of Bedrock on Weathering

The resistance of minerals to weathering depends both on their chemistry and mineralogy; for example, low temperature iron magnesium silicates such as hornblende are generally more stable during weathering than high temperature phases

TABLE III

Mean composition of rainwater and river water (from Garrels and Mackenzie, 1973).

	Rainwater ($m\ mol\ l^{-1}$ except TDS)	River water ($m\ mol\ l^{-1}$ except TDS)
Cl	0.107	0.22
Na^+	0.086	0.27
Mg^{2+}	0.011	0.17
SO^{2-}	0.006	0.12
K^+	0.008	0.06
Ca^{2+}	0.002	0.38
HCO_3^-	0.002	0.96
TDS (in $mg\ l^{-1}$)	7.13	130
Ionic strength	0.0001	0.002

such as olivine. Most minerals are soluble to some degree in surface conditions. Some, such as calcite, dissolve readily while other minerals, including the most abundant alumino-silicates, are only partially soluble and iteract with water to produce dissolved and solid phases.

Surface waters in areas of calcareous rocks - particularly sediments - contain more dissolved solids and are mainly Ca^{2+} - HCO_3^- waters with a high pH and are termed hard waters. In contrast, waters on igneous and metamorphic rocks contain lower quantities of dissolved solids, have a lower pH and contain different proportions of cations depending on bedrock chemistry and mineralogy. These waters are termed soft. The main properties of hard and soft waters are given in Table IV.

TABLE IV

The chemical properties of hard and soft water.

Property	Igneous Waters (Soft)	Calcareous Waters (Hard)
TDS	low	high
pH	6-8	7-9
Cations	Na^+, K^+, Mg^{2+}, etc.	Ca^{2+}, Mg^{2+}
Anions	HCO_3^-, H_4SiO_4	HCO_3^-
Weathered solids	Clay minerals	None

2.5. Chemical Elements in the Surface Environment and Factors Influencing Redistribution

2.5.1. Rock and Mineral Particles

Elements bound in resistate minerals are not chemically available and therefore unlikely to be of significance to health. However, particles of minerals of fibrous habit such as the asbestiform hornblendes may be associated with dust-related disease because of their physical properties.

2.5.2. Solutions

Elements dissolved in solution are present mainly as ions (electrolyte solution) and neutral molecules (non-electrolytes) of organic or inorganic compounds. The extent to which material dissolves in electrolyte form depends on the degree of ionic bonding. The chemical behaviour of dissolved species is determined by the effective concentration (or activity) of their ions.

The redistribution of elements amongst the different

weathering products is to a great extent determined by solution composition, with Eh and pH being of particular importance in mineral dissolution and precipitation.

2.5.3. Eh and pH

The Eh, or redox potential, of a solution is a measure of its oxidising capacity or its reducing capacity. By convention, positive values of Eh indicate oxidising conditions and negative values reducing conditions. It is particularly important in reactions involving S and such transition elements as Fe and Mn, which exist in different oxidation states in the normal range of surface conditions.

The pH of a solution, which is a measure of the H^+ ion content is important in controlling mineral dissolution and precipitation reactions of the major anionic species of the Earth's crust. It also affects such processes as ion exchange and complexing.

The chemistry of many elements in surface environments may be represented in diagrams which have Eh and pH as axes and which are standardised for temperature, pressure and activity (effective concentration) of major dissolved species. The techniques involved in construction of such diagrams are described in detail by Garrels and Christ (1965).

Although Eh/pH diagrams are of value in summarising the environmental conditions which cause dissolution and precipitation of different elements, they should nevertheless be used with caution since in many natural situations the controls on redox states and on dissolution-precipitation reactions are generally kinetic rather than equilibrium factors. For example, Eh/pH conditions may indicate that Fe^{2+} is the stable species although solid Fe_2O_3 may persist indefinitely if the rate of dissolution is slow.

Apart from kinetic effects, the assumptions used in construction of Eh/pH diagrams may not be valid; the thermodynamic data on which they depend often refer to pure substances, whereas many minerals have non-stoichiometric and/or variable compositions.

Finally, the diagrams are valid only for particular conditions at 25 °C, 1 atmosphere and specified anion concentrations, and any variations, particularly in anionic composition may affect predicted stability relationships. The assumption that simple mineral-solution equilibria predominate is reasonable for the major elements in the surface environment. The behaviour of many trace elements is more complex, however, and is also determined by co-precipitation and surface effects and interactions with organic phases.

2.5.4. Colloids and Particles of Secondary Phases

In natural waters both organic and inorganic compounds may occur as colloids which comprise very small particles of $1-10^{-4}$ μm in size.

The formation (peptisation) and flocculation of colloids has been intensively studied because of the role of colloids in transporting large quantities of metals in solutions for which true solubility would otherwise be low. Some colloids are exceptionally stable and such particles may remain in suspension indefinitely. The most important property of colloids is their surface charge which affects their interactions with dissolved ions. The process of ion exchange is important in the transport and redistribution of elements. Ions are held on colloids by electrostatic forces which range from weak to strong depending on the surface charge characteristics of the colloid and the ion. In general, however, the attachment of adsorbed ions is sufficiently weak to enable them to be replaced easily by other ions.

Another important process by which elements become fixed on colloids and particles is complexing whereby one or more central atoms or ions (usually metals) are attached to a number of ligands (ions or molecules).

Ligands are generally molecules containing atoms of electronegative elements with an unshared pair of electrons (C, N, O, S, F, Cl, Br, and I). Molecules containing more than one atom capable of donating a pair of electrons are known as chelating agents and include many natural compounds which are important in life processes. In natural systems both solid and aqueous complexes are common and complexes share the same range of solubility and stability as ionic and covalent compounds. In soft water from areas of crystalline bedrock and in soils generally, organic complexes (mostly as colloids) are likely to be of overwhelming importance in binding metals particularly those of the first row transition series. The organic complexes consist mainly of a complicated, poorly defined group of humic compounds which comprise fulvic and humic acids and an insoluble humic fraction. These substances have the capacity to complex considerable quantities of metal ions.

2.6. Uranium and Daughters

Uranium and its decay products are of special significance in the surface environment because of their chemical toxicity as well as their emission of ionising radiation. Uranium is highly mobile as the uranyl ion and forms carbonate and phosphate complexes in surface- and groundwater, especially where the pH is greater than 7.5. It is removed from aqueous solution under reducing conditions mainly by complexing with humic organic matter and iron oxides. For example, peat is known to have been enriched by a factor of 10,000 from surface water of average U content.

The daugther product, radon, is a well-known health hazard in ill-ventillated underground workings mainly due to the alpha-particle emissions of its solid decay products ^{218}Po, ^{214}Po, and ^{210}Po. Risks associated with exposure to radon in the mining of U are now well understood and controlled by adequate mine ventilation. A new hazard, however, has recently become apparent due to radon build-up in dwelling houses constructed of natural or fabricated materials of higher than normal radioactivity. This, combined with low ventilation conditions induced by a high degree of thermal insulation, can permit radon to reach dangerous levels.

2.7. Regional Geochemical Maps

Trace element levels cannot be predicted using a geological map which shows the distribution of different rock types. It is particularly difficult to deduce concentrations of elements over metamorphic rocks and certain types of sedimentary sequences, while levels of trace element over the same rock type may vary widely from average values. For example, values of Mo from 1 to 300 ppm and of U from 1 to 46 ppm have been reported over granites for which the average values are 2 and 3 respectively.

Ideally, trace element maps for application to agriculture or human health investigations should be based on systematic analyses of soil or vegetation samples. In Britain, such an approach has proved impracticable, because of the cost and time required to achieve an adequate sample density particularly taking account of seasonal variation. Information is available for parts of Scotland but in England, Wales and Northern Ireland there are few systematic data on either total or 'available' levels of trace elements

There are even fewer published data on regional variations of the levels of trace elements in pasture herbage and food crops. Some surveys of trace elements and heavy metal contaminants in food for human consumption based on random samples, and the analysis of total diets, have been published (Hamilton and Minski, 1972; Hubbard and Lindsay, 1975) and information on the intake of Hg, Pb, Cd, and As has been assessed. Surveys of this type do not, however, provide local information which enables epidemiological or public health studies to be made.

Metals in water supplies have also been monitored, both at abstraction points and in the household, in order to meet standards set by the European Economic Community for the quality of potable waters. Concentrations of metals in surface waters may vary appreciably on both a diurnal and seasonal basis, in relation to such factors as rainfall, and the movement of air masses over industrial areas. Further, it is difficult to compare the results of surveys carried out in different regions and at different times, particularly since they are often based on different sampling and analytical procedures.

Hence, available information on trace elements, including heavy metals in soils, vegetation, food and water, is generally inadequate for regional studies of the health of livestock or human populations.

The requirement for systematic data on trace element levels over the United Kingdom has been met, in part, by geochemical reconnaissance surveys, carried out by the Applied Geochemistry Research Group (AGRG) of Imperial College, and the Institute of Geological Sciences (IGS); the data obtained by these surveys are available in published atlases and in computer-readable form. (Figure 1).

The geochemical surveys are based mainly on the systematic collection and analysis of stream sediment

Fig. 1. Map showing distribution of molybdenum in stream sediments in England and Wales (compiled by the Applied Geochemistry Research Group as part of the Wolfson Geochemical Atlas of England and Wales, Webb et al., 1978).

samples. Each sample of sediment approximates to a composite sample of the erosion products of rock, overburden and soil upstream from the sampling point, and hence reflects the mean concentration of elements in the catchment area (Hawkes and Webb, 1962). Water samples are also collected by the IGS for elements such as U and Zn which tend to be relatively soluble in surface waters, while measurements such as pH, conductivity, total dissolved carbonate, F and total gamma activity are made in the field or field laboratory. The samples of stream sediments are analysed using a low-cost high-productivity method, such as direct reading emission or induction-coupled plasma spectrometry for up to 30-35 chemical elements. Large quantities of data are generated by the geochemical surveys and automated methods of data handling and cartography have been developed to prepare summary statistics and maps of the distribution of elements or element associations (Howarth and Martin, 1979).

A geochemical survey was conducted by the Imperial College in 1967 over the 5000 miles^2 of Northern Ireland and a geochemical atlas of the province including maps for 20 elements was published at a scale of 1 inch : 10 miles (Webb et al., 1973) A survey over the 64,000 miles^2 of England and Wales was carried out in 1969, and involved the collection of nearly 50,000 stream sediment samples over a ten week period. Large scale maps (1 inch : 4 miles) were placed on open-file at Imperial College in 1972-73. The Wolfson Geochemical Atlas of England and Wales at a scale of 1 : 2 million includes maps for 21 elements (Webb et al., 1978). The data for each element are presented in maps prepared using a local moving-average method of data smoothing to reduce 'noise' due to sampling and analytical error; this procedure also removes small-scale geochemical features that would relate to a specific farm or field, and is thus most appropriate for indicating broad-scale regional geochemical variation. Plots based on both empirical and percentile class intervals are included in the Wolfson Atlas, in addition to maps showing the distribution of multi-element associations (eg. Cu-Pb-Zn), by using a combination of 2 or 3 colours, each representing the distribution of an individual element.

The IGS survey is designed to produce larger (1 : 250,000) scale maps, with point-source data, and is based on specially designed sampling and analytical procedures (Plant, 1971) with stringent monitoring of procedural error (Plant et al., 1975). The distribution of each element is shown on separate maps by lines proportional to the element concentration. The geochemical data are plotted in black over a modern compilation of the geology prepared by automated methods as a single colour plot. In addition contour maps and geochemical 'landscapes' prepared by the SACM package at an approximate scale of 1 : 625,000 are included in the most recent atlases to indicate broad-scale geochemical trends. All the data from the IGS programme are also available on magnetic tape through the National Geochemical Data Bank. Geochemical atlases have been published for Shetland, Orkney, South Orkney and Caithness (IGS 1978a, b, 1979), Sutherland (IGS, 1980) and Hebrides (IGS, 1981). Geochemical data in computer-readable form are also available for most of the northern Highlands of Scotland and work on southern Scotland and the English Lake District has commenced. It is intended to extend the programme over the rest of Scotland and England and Wales.

A summary of chemical elements available in the published geochemical atlases is given in Table V. These geochemical data, generated by AGRG and IGS, may be combined with epidemiological data, and the possibility of interrelating datasets using interactive graphical devices

TABLE V

Summary of chemical elements available in geochemical atlases of the United Kingdom

	Published Atlases	Elements
Applied Geochemistry Research Group Imperial College	Northern Ireland	Al, As, Ba, Ca, Cr, Co, Cu, Ga, Fe, Pb, Mg, Mn, Mo, Ni, K, Sc, Si, Sr, V, Zn.
	England and Wales	Al, As, Ba, Cd, Ca, Cr, Co, Cu, Ga, Fe, Pb, Li, Mn, Mo, Ni, K, Sc, Sr, Sn, V, Zn.
	Shetland	Ba, Be, B, Cr, Co, Cu, Fe_2O_3, Pb, Mn, Mo, Ni, U, V, Zn, Zr.
	Orkney	Ba, Be, B, Cr, Co, Cu, Fe_2O_3, Pb, Mn, Mo, Ni, U, V, Zn, Zr.
	South Orkney and Caitness	Ba, Be, B, Cr, Co, Cu, Fe_2O_3, Pb, Mn, Mo, Ni, Sr (partial data), TiO_2 (partial data), U, V, Zn, Zr.
	Atlases on Open File	
Institute of Geological Sciences	Sutherland	Be, B, Cr, Co, Cu, Fe_2O_3, Pb, Mn, Mo, Ni, U, V, Zn, Zr.
	Lewis/Little Minch	Ba, Be, Bi, B, CaO, Cr, Co, Cu, Fe, K_2O, La, Pb, Li, MgO, Mn, Mo, Ni, Sr, TiO_2, U, V, Y, Zn, Zr.
	Great Glen	Ba, Be, Bi, B, CaO, Cr, Co, Cu, Fe, K_2O, La, Pb, Li, MgO, Mn, Mo, Ni, Sr, TiO_2, U, V, Y, Zn, Zr.
	Provisional maps available for purchase through NGDB	
	Argyll	Ba, Be, Bi, B, CaO, Cr, Co, Cu, Fe, K_2O, La, Pb, Li, MgO, Mn, Mo, Ni, Sr, TiO_2, U, V, Y, Zn, Zr.
	Moray/Buchan	F.

may facilitate the formulation of new hypothesis on relationships between geochemistry and health.

2.8. Regional Geochemistry of Britain

Britain can be conveniently divided into three physiographic/ geological regions (Figure 1). These are:

(a) The Precambrian crystalline basement rocks of northern Scotland.

(b) The Palaeozoic shale/greywacke sequences, containing granite intrusions, of southern Scotland, the Lake District, Wales and south-west England; these areas contain most of the important metallogneic provinces in Britain.

(c) The Devonian-Tertiary sedimentary cover succession of England and South Wales.

Regions 1 and 2 are upland areas where rainfall is high and impervious crystalline rocks give rise to rapid rates of run-off. Water-rock contact times are low, and are associated with low concentrations of the major cations and anions reflected by exceptionally low conductivity values; (IGS 1978, 1980, and 1981). Concentrations of H^+ ions generally exceed Ca^{2+} ions and surface conditions are predominantly acid with large areas of peat bogs and/or acid soils. The acidity may be further increased by air masses containing acid gases (e.g. SO_2) from industrial areas. In such surface waters increased levels of heavy metals can be mobilised as a result of enhanced acidity and the presence of colloidal and dissolved organic substances capable of adsorption or forming complexes. Moreover, soft water containing organic acids may dissolve quantities of heavy metals from unplasticised PVC or metal water-distribution systems producing high levels of heavy metals in drinking water. Increased concentrations of heavy metals in soft water should be considered in addition to the low levels of major ions such as Mg in determining the importance of the water hardness factor in epidemiology. These environments are generally deficient in the major cations and anions, including the nutrient phosphate ion, although some adsorption, and accumulation, may occur on clay and soil particles in river valleys. In such acid reducing conditions, Fe and Mn are soluble. However, precipitation can take place near to the soil surface as iron pans. Also in the case of streams in equilibrium with atmospheric oxygen, the formation of insoluble, hydrous manganic and ferric oxides can take place.

In contrast to these conditions, large areas of England (Region 3) consist of agricultural soils underlain by permeable sedimentary rocks. At the surface, waters are predominantly of the Ca^{2+} - HCO_3 type. Eh and pH and levels of the major cations (Ca^{2+}, Na^{+}, Mg^{2+}, and K^{+}) and anions are generally higher with a consequent increase in water hardness. The content of organic acids is low, since they are precipitated under such conditions. Mobilisation of heavy metals by complexation and adsorption on colloidal organic matter is also correspondingly reduced.

In deep aquifers such waters evolve into chloride-rich brines with concentrations of Na^{+}, K^{+}, Ca^{2+}, and Mg^{2+} increased by at least an order of magnitude. Increased concentrations of such elements as Sn, Zn, Ba, and Pb are stabilised by chloride complexes. Such deep waters may also contain large quantities of F, which is normally buffered at low levels in the surface environment by precipitation of CaF_2. In relatively deep waters (which provide the water supply in many parts of England) the controls on trace element concentrations are chemical equilibria, speciation and the solubility of metals.

Within each of the three regions, and particularly

Region 1 and 2, bedrock geochemistry will be an important factor in determining the surface geochemistry.

2.8.1. Bedrock Geochemistry

Region 1. Precambrian crystalline basement rocks of northern Scotland.

As a result of its geology and climate, levels of Ca, Na, Mg, K, and P tend to be low in the surface environment of northern Scotland, except in areas underlain by sediment and basic rocks. Most of the region, including much of the area underlain by Old Red Sandstone sediments, is also low in Fe, Mn, Cr, Ni, Zn, Cu, Co, and V. The mobility of these elements may also be limited by high pH and carbonate concentrations which occur in areas underlain by sediments. The same conditions increase the availability of Mo, Se and U in dissolved and/or colloidal phases. Elsewhere availability of the first row transition elements may be enhanced by adsorption or complexing with colloidal oxides or organic matter. The high biological accessibility of these materials indicates that potentially toxic levels of trace elements may occur in an environment generally low in essential elements. The availability of F in the predominantly low Ca regime is likely to be increased, and in some cases (e.g. the high Mo levels in Caithness) toxic levels of an element may exacerbate a deficiency condition, as Mo does by blocking the utilisation of Cu in ruminants. These patterns suggest that northern Scotland may be an area in which studies of environmental geochemistry and health might prove particularly fruitful.

Region 2. Palaeozoic shale/greywacke belts with granite intrusions of England and Wales.

As in northern Scotland, this region is likely to be characterised by deficiencies of major elements as a result of bedrock composition (low Ca), low mineral solubility/high degree of leaching (K), and the formation of insoluble hydroxides of Fe and Mn. Levels of most essential trace elements are likely to be reasonably adequate, however, except over areas underlain by granite.

Region 3. Old Red Sandstone to Tertiary cover sediments of England and Wales.

The levels of both major and trace elements in Region 3 are closely related to lithology. Areas underlain by sandstones and limestones contain low or very low levels of most of the essential trace elements; limestones are also low in K and Fe and sandstones in Ca. Low levels of trace elements in sandstones are sometimes enhanced by surface conditions which are frequently acid with excessive leaching. Potentially toxic amounts of Mo occur in black shales of Jurassic and Carboniferous age and high levels of Pb, Zn, and Cd in the region are associated with mineralisation, and old mineral workings.

2.9. Trace Elements in Soils

The main sources of trace elements in soils are the parent materials from which they are derived. In Britain this is weathered bedrock or overburden transported by wind, water or glaciation. Transported material is, in general, of relatively local origin, though there are some notable exceptions. Thus, as indicated above, soils developed from acid igneous rocks such as rhyolites and granites and from coarse grained sedimentary rocks usually contain smaller amounts of the nutritionally essential trace elements than those derived from basic igneous rocks or fine-grained sediments.

For example, the average copper content of British surface soils has been recorded as 20 ppm. However, the range in apparently uncomtaminated soils is wide and largely reflects the nature of the parent material. Data for a large number of surface soils from England and Wales show total Cu levels ranging from as little as 2 ppm in soils derived from Pleistocene sands to 60 ppm or more in those from some marls, shales and clays (Thornton and Webb, 1975).

Levels of trace elements found in normal and metal-rich soils are given in Table VI. For example, calcareous soils derived from interbedded shales and limestones of the Lower Lias formation in Somerset, containing 20 ppm Mo or more, have been associated with molydenosis in grazing cattle. Soils derived from ultra-basic rocks containing nickel-rich ferromagnesium minerals provide a further example of a natural source of metal excess. In parts of Scotland such soils under poor drainage conditions may give rise to nickel toxicity in cereal and other crops.

Soils in mineralised areas in Britain frequently contain large amounts of one or more of the elements Cu, Pb, Zn, Cd, and As. These soils may be developed over or in the dispersion halo around ore-bodies and mineral veins or from transported overburden containing mineralised materials. The majority of mineralised areas in Britain have been worked and surface soils are usually contaminated to varying degrees by mining, processing and smelting operations. It is often difficult to distinguish whether anomalous concentrations of metals in the soils of these areas are due to natural weathering of underlying mineralised materials or to inputs from mining activities.

Where soils are developed *in situ* from the underlying strata, agricultural problems may be directly related to the

TABLE VI

Trace elements in soils derived from normal and geochemically anomalous parent materials.

Normal ranges in soil mg kg^{-1}		Metal-rich soils mg kg^{-1}	Sources	Possible effects
As	<5-40	up to 2500	Mineralisation	Toxicity in plants and livestock
		up to 250	Metamorphosed rocks around Dartmoor	Excess in food crops
Cd	<1-2	up to 30	Mineralisation	Excess in food crops
		up to 20	Carboniferous black shales	
Cu	2-60	up to 2000	Mineralisation	Toxicity in cereal crops
Mo	<1-5	10-100	Marine black shale of varying age	Molybdenosis or molybdenum induced hypocuprosis in cattle
Ni	2-100	up to 8000	Ultrabasic rocks in Scotland	Toxicity in cereal and other crops
Pb	10-150	1% or more	Mineralisation	Toxicity in livestock; excess in food crops
Se	<1-2	up to 7	Marine black shales in England and Wales	No effect
		up to 500	Namurian shales in Ireland	Chronic selenosis in horses and cattle
Zn	25-200	1% or more	Mineralisation	Toxicity in cereal crops

bedrock geochemistry. Where parent materials have been mixed or redistributed by glacial activity or fluvial transport, the trace element content of the underlying rock may be completely masked or modified or its effect smeared in the direction of ice movement or water flow.

The influence of parent material on both the total content and form of trace elements in the soil is modified to varying degrees by the processes of soil formation or pedogenesis, which may lead to the mobilisation and redistribution of trace elements both within the soil profile and between neighbouring soil types. The extent to which the parent material-topsoil relationship is affected depends on both the ease of weathering of the primary minerals and the age of the soil. In the relatively young soils over much of Great Britain, the parent material remains the dominant factor in determining the soil trace element status. However, gleying, leaching, surface organic matter accumulation, podzolisation, together with soil properties such as reaction (pH) and redox potential (Eh) may affect the distribution, form and mobility of trace elements in the soil and their availability to plants. For example, trace elements are frequently depleted in the surface horizons of podzols with subsequent enrichment in the B horizon.

Information on trace elements in soils is available for parts of Scotland from the analysis of samples taken both for soil survey and advisory work. However, there are few systematic data on either the total or available levels of trace elements in soils in England and Wales. Numerous ad hoc studies have been made by specialist department of the Ministry of Agriculture, Fisheries and Food, the research institutes and universities, though to date there has been little, if any, effort to collate results. However, the Soil Survey of England and Wales has published data for trace elements in soils of Sheet SK17, Tideswell in Derbyshire, Sheet TR04, Ashford in Kent, and Sheet TF16, Woodhall Spa in Lincolnshire.

Studies by the Agricultural Development and Advisory Service and the Soil Survey based on grid sampling in the Halkyn area of North Wales, and on traverses in the Hayle area of Cornwall will provide useful data on the distribution of heavy metals in the soils of these old mining areas.

One of the most comprehensive collections of data showing mainly normal or background levels of both total and 'available' trace elements in soils is that compiled in connection with the 'Survey of Fertilizer Practice' by the Agricultural Development and Advisory Service, based on a random sample of approximately 1500 farms (Archer, 1980).

2.10. Metal Pollution

The interface between natural geochemical sources and anthropogenic inputs of chemical elements into the soil is often difficult to discern. The wide range in natural background levels of metals in soil due to the chemical nature of soil parent material is important in the interpretation of pollution surveys; similarly man-made pollutants may increase the harmful effect of geochemical sources of metal.

2.11. References

Archer, F. C.: 1980, 'Trace Elements in Soils in England and Wales', in Inorganic Pollution and Agriculture, H.M.S.O., London.

Fyffe, W. S.: 1974, Geochemistry, Clarendon Press, Oxford, Oxford Chemistry Series, 16.

Garrels, R. M. and Christ, C. L.: 1965, Solutions, Minerals and Equilibria, Freeman Cooper, 2nd edit. 450 pp.

Hamilton, E. I. and Minski, M. J.: 1972, 'Abundance of the Chemical Elements in Man's Diet and Possible Relations with Environmental Factors', Sci. Tot. Envir. 1, 375-94.

Hawkes, M. E. and Webb, J. S.: 1962, Geochemistry in Mineral Exploration, Harper and Row, New York, 415 pp.

Howarth, R. J. and Martin, L.: 1979, 'Computer-based Techniques in the Compilation, Mapping and Interpretation of Exploration Geochemical Data', in Hood, P. J. (ed.), Geophysics and Geochemistry in the Search for Metallic Ores, Geol. Surv. Can. Econ. Geol. Rep. 31, pp. 545-74.

Hubbard. A. W. and Lindsay, D. G.: 1975, 'Control

Surveillance of the Contamination of Food by Heavy Metals in the United Kingdom', Symp. Proc. Int. Conf. Heavy Metals in the Environment, Toronto (ed. T.C. Hutchinson et al.), 1, 163-72.

Institute of Geological Sciences: 1978a, 1978b, 1979, 1980, 1981, Geochemical Atlas of Great Britain: Inst. Geol. Sci. London.

Levinson, A. A.: 1974, Introduction to Exploration Geochemistry, Applied Publishing Ltd., Calgary, 612 pp.

Plant, J.: 1971, 'Orientation Studies on Stream Sediment Sampling for a Regional Geochemical Survey in Northern Scotland', Trans. Instn. Min. Metall. B80, 324-44.

Plant, J., Jeffrey, K., Gill, E., and Fage, C.: 1975, 'The Systematic Determination of Accuracy and Precision in Geochemical Exploration Data', J. Geochem. Explor. 4, 467-86.

Thornton, I. and Webb, J. S.: 1975, 'Distribution and Origin of Copper Deficient and Molybdeniferous Soils in the United Kingdom', Proc. Copper in Farming Symposium. London: Copper Development Ass.

Underwood, E. J.: 1977, Trace Elements in Human and Animal Nutrition, 4th ed. Academic Press, London and New York.

Webb, J. S., Thornton, I., Thompson, M., Howarth, R. J. and Lowenstein, P. L.: 1978, The Wolfson Geochemical Atlas of England and Wales, Oxford University Press, Oxford.

Webb, J. S., Nichol, I., Foster, R., Lowenstein, P. L., and Howarth, R. J.: 1973, Provisional Geochemical Atlas of Northern Ireland. Appl. Geochem. Res. Group Tech. Commun. 60, Imperial College, London.

3. PLANT-SOIL PROCESSES

3.1. Summary

The major elements, N, P, K, Ca, Mg, and S and minor elements, Cu, Fe, Zn, Mn, B, Mo, and Cl are essential for plant growth, though some may be phytotoxic, such as Cu, Zn, and Mo, if present in large amounts in the soil. Several non-essential elements, including Ni, Cd, Hg, Se, and Pb, may occur naturally or due to contamination and may be toxic to plants or to animals consuming them. The needs of plants for N, P, K, Ca, and Mg are well understood and most may be predicted by soil analysis. There is, however, still a need for a better understanding of the slow releases of P and K from many soils. The behaviour and availability of trace elements is much less well understood but are of major relevance to relationships between geochemistry and plant health. The association between region or soil type and trace element status is less clear than that for major elements, and broadscale data published in the form of geochemical atlases is of value in detecting regional patterns of trace element distribution.

Plant factors influencing the availability of elements in the soil comprise uptake processes, element interaction, growth rate, root factors, species and genotype effects, composition of plant parts, trace element function in plants and crop offtake. Uptake of the majority of trace elements is 'active' or at least metabolically controlled. Uptake by roots is in part related to the concentration of elements in the soil solution, though there is little basic information for trace elements, with the exception perhaps of Cu and Zn. All transition metals form complexes with soluble organic ligands in soils and there is little quantitative evidence on the uptake of these complexes in relation to their concentration. Root hairs and mycorrhizas associated with roots also influence rates of uptake but there is little specific information for trace elements. Metal ion uptake can be inhibited by major cations in the soil and antagonistic effects between trace metals are recognised. Soil pH has a major influence on trace element uptake and also on the formation of metal complexes. The concentrations of trace elements in plant tissue may be diluted by growth, and metal concentrations in pastures are largest in winter and spring. Root factors influencing uptake by modifying the root zone include pH change, release of root exudates capable of complexing trace metals and of compounds able to reduce Mn and Fe oxides, and microbiological activity.

Different plant species and cultivars vary widely in their ability to grow on deficient or toxic soils, and at the same time may take up varying amounts of specific elements. The concentration of elements in plants is related to the stage of growth and is usually greatest in young plants. Breeding plants resistant to mineral stress has been successful, though the mechanisms of susceptibility or tolerance are not fully understood. The selection of cultivars for the composition of their edible parts has received little attention and could be of value for metal contaminated land. The concentration of trace elements varies between plant parts and is related to their mobility in the plant, varying with the element and the level of supply. The function of essential trace metals in plants is concerned with specific enzymes; only a fraction of the total metal in the plant is involved, and for some elements the function is still not known.

The concept of availability of elements in the soil to plants is difficult to define and complex, depending on the equilibrium concentration of the ion in the soil solution, the chemical forms in which it is present, and the rate and degree to which the concentration is maintained near an absorbing root. The relationship between total amounts of elements present and plant health is not fully understood and empirical chemical extractants have been used with limited success to diagnose 'plant-available' fractions. Availability to plant depends on those fractions of elements present in the soil solution. This is strongly influenced by soil pH, oxidation-reduction processes of particular importance to elements such as Fe and Mn of variable valency, physico-chemical sorption on to the surface of colloids such as Fe and Al oxides and clays, and the degree to which elements form complexes with organic ligands. Ions move by mass flow and/or diffusion to roots. The speciation or chemical forms of an element in the solution and solid phases of the soil are a function of complicated physico-chemical relationships. Computer simulations have been and continue to be developed to define chemical species present in the soil system.

Grassland herbage production is rarely affected by trace element deficiencies. However, its nutritional quality for livestock is associated with deficiencies of Co, Cu, Mn, I, and Se (and P and Mg) and an excess of Mo and Se. Boron deficiency is recognised in Britain particularly in sugar beet. Manganese deficiency is found in many crops including cereals, particularly in soils of high organic content and high pH. It is controlled with leaf sprays and is diagnosed by leaf symptoms supported by plant analysis. Manganese toxicity is found on acid soils and is cured by liming. Zinc deficiency has yet to be recognised in crops in Britain with the exception of fruit, though toxicity due to excess is sometimes found in metalliferous mining areas and with the disposal of sewage sludge and other wastes. Copper deficiency is not uncommon in cereal crops, particularly on calcareous and highly organic or very sandy soils, and responds to soil dressing of Cu; soil analysis has proved a useful diagnostic aid. Copper toxicity is an uncommon problem.

Information indicating the likelihood of trace element problems occurring in an area is desirable. This is best derived from clinical evidence of disease and from field trials with extrapolation into similar areas. The use of soil parent material for this extrapolation can be useful but is subject to limitations, as it may vary in chemical composition and may be modified by soil forming processes. Geochemical surveys using stream sediment sampling have proved useful in denoting the total levels of trace elements in soils. Ideally soils should be analysed for total element content and that extracted by an accepted extracting solution, though the latter is complicated as any one extractant may not be effective as a means of diagnosing potentially deficient or excessive amounts of all the different elements to be tested. Statistically based sampling techniques and data handling procedures are essential for the assemblage of reliable trace element information for soils. High density sampling is required especially in those areas indicated as containing exceptional levels of more than one element by stream sediment surveys.

3.2. Introduction

Elements taken up by plants from the soil can, for convenience, be divided into three groups:

(a) Those required as major or minor essential elements; N, P, K, Ca, Mg, S, Cu, Fe, Zn, Mn, B, Mo, and Cl. There are also a small number of 'functional' elements which aid growth but are not essential: Na, Co, Si.

(b) Elements found in the soil, either naturally or following artificial contamination, at such levels that they are phytotoxic. Some of these overlap with (a), for example Cu, Zn, and Mo. In addition there are several non-essential but potentially toxic heavy metals such as Ni, Cd, Hg, and Se.

(c) Elements whose supply is of little or no importance for plant growth, but whose concentration in the plant is important for the health of animals or humans consuming the plant, e.g. Co, Se, I, Cr, and As.

3.3. Major Elements N, P, K, Ca, and Mg

Geochemical factors affect crop growth very strongly through the major element deficiencies, and it is only the ease and frequency with which we correct these imbalances with artificial fertilizers which often makes us ignore this as an aspect of geochemistry. If not corrected, such major element deficiencies may reduce yields in most of the farmed land area of the world; for example, phosphate deficiency initially prevented useful farming in large areas of Australia, and currently limits it in the Cerrado regions of Brazil. In total, some 400,000 tons of P_2O_5 and of K_2O are used in the U.K. every year to correct soil deficiencies. The very large amounts of N used every year are not further mentioned here, because the need for this is controlled more by past cropping and the soil texture rather than by more strictly geochemical processes.

Those deficiencies, which result from the composition of the parent rock or from the pedogenic processes, occur in most soils, but repeated applications of fertilisers have mitigated the problem in many places. It has been estimated, for example, that 45% of all P in British agricultural soils has been added artificially. There are, however, quite marked differences between soil types which are still of practical importance. There are well-recognised deficiencies associated with parent material and soil type, such as K deficiency on chalk rendzinas and light sands, and P deficiency on some of the Chalky Boulder Clays. Magnesium deficiency is recognised and corrected on about 50,000 ha, mainly of light land. The needs for Ca as an essential element is always fully met, but Ca is also needed as the dominant cation in soils other than highly acid ones, and very large amounts of lime (ca. 3,000,000 tons yr^{-1}) are applied in Britain to correct acidity. This need is strongly dependent upon soil texture, farming system and the calcium carbonate naturally present in the soil. Correction of low pH prevents toxicity of Fe and Mn and of other metals if they are present in excessive amounts. It can also lead to enhanced uptake of Mo, suppression of Co uptake, or induced deficiency of Fe or Mn.

A number of soils with high clay content, such as those developed on the Chalky Boulder Clay, release K so readily that no applications are necessary unless very demanding crops are grown. For example, work at Rothamsted has shown that very intense cropping with grass in pots can remove K up to 2.2% of the total weight of Evesham series soil. This release is of greatest importance in low-intensity use, such as forestry or rough pasture, where the value of the product does not allow heavy fertilizing.

The needs for P, K, Ca, and Mg are now well understood and can be predicted satisfactorily on the basis of soil analysis (MAFF, 1979). They are usually associated with well known soil series, and while the history of the individual field is of great importance, it is possible to make useful generalisations. This topic has been so thoroughly researched in the past that there are few pressing problems at present, though minor improvements are always being made, and the steady increase in crop yields and nutrient off-takes require a continuing interest in these problems. A better understanding of the slow processes of release of P and K which occur in many soils would be desirable.

3.4. Other Elements

The behaviour and availability of the trace elements (essential or not) is much less well understood than that of the major nutrient elements, and consequently the problems associated with the former are now regarded as much more pertinent to geochemistry. Some of these trace elements are applied regularly as crop fertilisers, or as amendments essential to the health of stock. For others the interest has only developed very recently, based often on suggestions of a linkage with some aspect of health.

The association between region or soil type and trace element status is less clear than that for the major elements, except for particular areas with rather outstanding and unusual properties. Valuable broad data are now available in the Wolfson Geochemical Atlas of England

and Wales and in the new IGS stream sediment surveys. Such surveys have successfully detected major anomalies in the incidence of different elements in the pedosphere, a number of which have agricultural implications.

3.5. The Availability of Elements: Plant Factors

3.5.1. Uptake Processes

It seems certain that uptake of the majority of trace elements is 'active' or at least metabolically controlled (see Tinker, 1981). Most evidence to the contrary is from uptake into excised tissues which is not a good guide to the behaviour of intact plants. However, the cell walls contain carboxyl and hydroxyl groups which readily complex with transition metals, so that appreciable amounts of an ion may be bound passively in this way in the root, without passing the cell membrane.

There is currently controversy over how uptake by intact plant roots is related to solution concentration around them, because for some major ions (e.g. NO_3) the effect of concentration appears to be slight. Loneragan (1975) concluded that for Zn, B, Cu, Mn, and Fe, there was a linear relationship between uptake rate and concentration, up to values of the latter much above 10^{-6} M, which is a high concentration in the soil solution for these elements. However, there is little basic information on the relevant points for trace metals, and very few accurate and dependable whole-plant studies in high-volume flowing solution culture, which is the only dependable way of investigating uptake physiology of the whole plant. The work that has been done has tended to concentrate upon the well-known elements such as Cu and Zn, and little is known about the less important elements. The comparison of rates of uptake of trace metals with those of the major elements (see Tinker, 1981) indicates that the absorbing power for the former is relatively low. More work along these lines seems important.

Perhaps the greatest complication in uptake from soil is that all transition metals are able to form complexes with soluble organic ligands, e.g. in soil over 90% of Cu will normally be complexed. So far there is little quantitative evidence on the relationship between uptake rate and concentration of complexed species. There is some general evidence that the complex must split before uptake occurs, so that it is the free ion which is absorbed. The net effect of the complex-forming ability of an ion is thus complicated, involving complex formation with insoluble materials in the soil so that the ion is immobile; complex formation with soluble ligands in the soil solution so that it may move to the root surface; complex formation with root cell wall material, which may interfere with active uptake; and complex formation as part of the active uptake step. The interrelationship between this series of processes is probably the major reason why heavy metal uptake is so poorly understood at a fundamental level.

The complicating effects of root hairs and mycorrhizas cannot be ignored. Root hairs are likely to increase root uptake efficiency for all strongly sorbed elements with low diffusion coefficients in the soil, but very little work has been done on this for trace metals. Vesicular-arbuscular mycorrhizas absorb Zn and Cu and transfer it to the host plant, and are also involved in the Zn-P interaction in uptake. Recent work at Rothamsted has shown that Zn- and Cd-tolerant strains of the mycorrhizal fungus exist.

3.5.2. Element Interactions

When elements are supplied in the uncomplexed form in solution culture, increasing pH depresses uptake of Mo, but generally increases that of the metallic cations. However, the presence of complexing agents may reverse these effects, and in soil the final effect of pH on uptake of micronutrients is strongly conditioned by the interaction of trace element ions with soil colloid surfaces (see Loneragan, 1975). For this reason the uptake of Mo from soil increases with pH, but that of Zn, Mn, and Fe decreases, whereas uptake of Cu is less strongly affected. Boron deficiency is also enhanced at high pH.

The uptake of the metallic trace ions is inhibited by the major cations, especially by Ca. Many trace elements also interact, e.g. both Cu and Zn reduce the uptake of the other, and Zn affects Fe level in plants. There are many reports of the antagonistic effects of phosphate on uptake of Cu, Zn, and Fe, whereas it may enhance uptake of Mo. It is interesting that sulphate depresses uptake of Mo, and that there are suggestions of a Cu-Mo antagonism in plants, bearing in mind joint interactions with Cu in animal nutrition. The very similar ions selenate and sulphate, and arsenate and phosphate are antagonistic to each other in uptake, though the effects in soil may be less simple (Tinker, 1981). Interactions are thus frequent and important, but the understanding of the mechanisms involved, or even the predictions of their occurrence in field situations, is presently very poor.

3.5.3. Growth Rate

The internal concentration of a trace element in plants may depend upon the degree of dilution by growth, e.g. Pb content in pasture species is always largest in winter and spring, but decreases during active growth later in the summer (see Petersen, 1978). The supply of extra N to cereals accentuates the deficiency symptoms of Cu deficiency (MAFF, 1971), and it seems likely that this is due to the increased growth caused by the N.

3.5.4. Root Factors

Roots modify their environment quite extensively in many ways. The most important of these are pH change, exudation and microbiological activity in the rhizosphere.

Large pH changes are very well substantiated, up to at least 1 pH unit, depending upon the form of N nutrition which determines cation-anion balance. Root exudates certainly contain compounds capable of complexing trace metals, for example hydroxycarboxylic acids and amino acids, and the soil solution concentration of Cu may increase greatly when plants are grown in such a soil, probably due to complex formation. Roots also release compounds which can reduce ferric and manganese oxides to ferrous and manganeous salts respectively. There are likely to be differences in this reduction process between species and cultivars, and Fe-deficient plants excrete larger amounts of such reducing compounds than similar non-deficient plants.

3.5.5. Species and Genotype Effects

It has been known for a long time that cultivars (and of course plant species) vary considerably in their ability to grow on deficient or toxic soils. The most striking case of this is in the element accumulators, such as Astralagus species which contain Se, and Becium homblei which contains Cu in large amounts. Such classic instances do not occur in

Britain, but large accumulation of As was found in Agrostis tenuis, though it was not considered to represent a hazard.

Different species also react differently to varying availability of elements, thus the Cu, Mo, and Co content of clover appears to vary more widely than that of ryegrass (see MAFF, 1971). Burridge (1970) has discussed the variation of mineral compositions with species; in a comparison of many samples of ryegrass and clover from two sites, the only completely consistent result was that Mn was always largest in ryegrass, and Fe in clover; Co tended to be greatest in clover, and Cu in ryegrass. However, clover may contain more of Cu, Mn, and Zn when soil supply is large. Amongst vegetables, lettuce appears to accumulate Pb and Cd more than the average ones. The stage of growth is also important, because young plants typically contain larger concentrations of all elements than old ones. It is therefore important to compare plants at similar stages of development.

There is the possibility of breeding specifically for plants which are resistant to various forms of mineral nutrient stress and Al- and Fe-tolerant strains of cereals are in use in other countries. Indeed, testing for these properties may be essential, because potential susceptibility to deficiencies or toxicities may be missed in the normal breeding programme. There is now a wide range of information available on several elements. The mechanisms of susceptibility or resistance however are still not fully understood, and there may indeed be several of these. In heavy metal-tolerant lines of grasses, there is much evidence that they can retain a larger fraction of the element in the root, by binding it more firmly in the cell walls. Tolerant cultivars growing on heavy metal-contaminated land will therefore contain much larger concentrations of the metal in the roots.

It would be a serious matter if this were also the case for shoots, because such vegetation might more easily be dangerous to grazing stock. However, direct comparisons have shown that there is no consistent difference between tolerant and non-tolerant cultivars growing on the same soils and one tolerant line of Festuca rubra contained less Pb and Zn than a commercial variety (Johnson et al., 1976). There thus does not appear to be an enhanced hazard from using such tolerant cultivars, except possibly when they will grow on soil where non-tolerant cultivars would die. Rather little attention has been given to the possibility of selecting cultivars for the composition of their edible parts, but the possibilities are indicated by large differences in Cd concentration in different cultivars (see HMSO, 1980).

3.5.6. Composition of Plant Parts

The concentration of trace elements always varies between plant parts. Fe tends to be the most variable, and Co, Cu, and Mn the least; presumably the variability of Fe is due in part to its lack of mobility in the plant. Other ions are re-translocated fairly freely between different plant parts (see Tinker, 1981), but the ease with which this occurs, and the fraction present in the different plant parts, varies greatly with the element and with the level of supply. In general, the reproductive parts and the seed vary least in composition. However, samples of grain from very different conditions can have a range of compositions, e.g. different authors have reported ranges from 26-50 ppm (Fe), 20-50 (Zn), 4-7 (Cu), 0.22-0.5 (Mo), 0.008-0.13 (Se), 0.009-0.1 (As), 0.009-0.25 (Cd), 0.0006-0.008 (Hg) and 0.005-0.03 (Co) (see Lag, 1978). These ranges are relatively much wider for the elements which are non-essential to plants. There are also significant differences between different plant parts. Roots, especially the fibrous roots, often contain larger concentrations of Pb than leaves, whereas for Cd the reverse may occur, but there is much variation, and generalisations may not be dependable.

3.5.7. Trace Metal Function in Plants

The need for most essential trace metals is connected with the function of specific enzymes, either as components of the enzyme molecule or as co-factors affecting its efficiency. For some, such as B, the function is still obscure. Only one enzyme, nitrate reductase, requires Mo; for others, such as Mn, many enzyme functions, and parts of the photosynthetic process, depend upon their presence. It is important that only a very small fraction of the total amount of metal in the plant is actually combined in the enzyme. The rest is in various forms of combination, including low molecular weight complexes in the plant fluids, and metal complexed on the plant cell walls.

3.5.8. Crop Offtake

The total amounts of trace elements in the surface soil are normally so much greater than the crop contents that no depletion in the simple sense could be expected, but appreciable fractions of the 'available' element may be removed (Table I). It is a little disturbing that most data on crop content are from much earlier work, with crops which now would be regarded as very poorly yielding, and it appears desirable to investigate the composition and total content of modern high-yielding crops.

TABLE I

Approximate weights of some trace elements and major elements removed by crops and in the soil (kg ha^{-1})

	Removed by average crop	Extractable by diagnostic reagent	Total content in soil
Cobalt	0.001	0.2 - 4	2 - 100
Molybdenum	0.01	0.002 - 1	0.5 - 10
Copper	0.1	1 - 20	2 - 100
Boron	0.2	1 - 5	4 - 100
Zinc	0.2	2 - 20	20 - 50
Manganese	0.5	10 - 200	100 - 10,000
Iron	0.5	10 - 200	2000 - 1,000,000
Magnesium	20	100 - 10,000	2000 - 1,000,000
Phosphorus	20	40 - 100	1000 - 10,000
Potassium	100	40 - 1000	5000 - 50,000

After Russel, E. W. (1973) Soil conditions and plant growth. 10th edition, p. 642.

3.6. The Availability of Elements in the Soil: Soil Factors

3.6.1. Quantity and Chemical Form

It is usually accepted that the total quantity of a trace element has little direct influence on whether plants growing in that soil show a deficiency, though there is usually a relationship when the amounts are extreme. This is due to the very diverse forms in which the elements can occur, from surface-sorbed material in rapid equilibrium with the soil solution to pieces of unweathered mineral. The range of total quantities is in Table I, (Mitchell and Burridge, 1979). Thus very marked deficiencies of Cu appear of sandy soils with very low total amounts, Mo-induced hypocuprosis occurs on soils with large amounts of Mo, and contaminated soils or mine spoil normally have large amounts of both total and extractable metal. Recent work at Rothamsted showed a surprisingly close relationship for several trace metals between total and extractable amounts, when the comparison was made within some individual soil series.

The problem raised by complex formations, pH effects and element antagonisms can be considered under the general heading of 'speciation' - the chemical form of the element in the solution and solid phases of the soil. The physico-chemical relationships are extremely complicated because of the large number of chemical species involved, and the overriding effects of pH and redox potential (Jones and Jarvis, 1981). Very broad generalisations can easily be made, but exact treatments are now becoming available in the form of computer solutions of the simultaneous equations which govern the system (Mattigod et al., 1981). These are a great basic advance, but should not be regarded as a simple solution to the question, because they require a very large amount of detailed information, and may not allow for the delayed equilibria and impure solid phases in soils.

Soil reaction is probably the single most important factor in trace element chemistry. The transition metals all form highly insoluble oxides or hydroxides which reduce the concentration of the free ion in the soil solution to minute levels at alkaline pH.

Oxidation-reduction processes are specially important for elements of variable valency; thus Mn is a difficult case, because its most stable oxide, MnO_2 is a quadrivalent form of Mn. In principle, Mn^{++} concentration in equilibrium with MnO_2 at pH 7.0 should be only $10^{-16.4}$ M, but it is greater than that in all analysed soils, which stresses the point that soil can never be regarded as being in complete physico-chemical equilibrium with single inorganic compounds.

Most elements undergo physico-chemical sorption on the soil colloid surfaces, e.g. the ion concentration of Cu and Zn in soils is well below that predicted from the solubility of the hydroxides, so that equilibrium is clearly maintained by sorption. Mo, B, As, and Se are normally present as anions and behave in different ways. The MoO_4^{-2} ion is the most common Mo species, and this adsorbs on soils in ways analogous to that of the phosphate ion. However, Mo adsorption increases with decreasing pH, so that Mo availability always increases with pH. The species of B present in the soil solution is mainly H_3BO_3 produced by the soil microbial population, and this is not generally in direct equilibrium with any soil mineral. However it may be sorbed on Fe and Al oxides and on clays. Both sorption processes are pH dependent, with maxima usually in the pH 7-8 range. Both selenates and selenites exist in soil, and the latter can be quite strongly sorbed on soil colloids. Arsenic is usually present as arsenate, which sorbs very much like phosphate.

Almost all elements we are concerned with will complex with organic ligands to some degree which thus influence plant nutrition. Unfortunately the ligands present in the natural soil solution are poorly defined, and will certainly vary between soils. For a 'fulvic acid' preparation, the order of complex stability was: Cu > Fe > Ni > Pb > Co > Ca > > Zn > Mn > Mg, but it varies with pH. This confirms that complex formation will be of major importance for Cu and Fe but less so for Zn and Mn. At present our understanding of the partition of a given ion between the different types of complex-forming species is poor. Such work as has been done is on 'humic' and 'fulvic' acids, which are to some degree artefacts following their extraction from soil. Work in Scotland has indicated that fulvic acid is closely similar in its properties to the organic material in the natural soil solution, but much further study of this topic seems to be needed.

3.6.2. Availability

There is no one value for the 'availability' of a given element in a given soil, and indeed the concept of 'availability' is difficult to define, and is now regarded rather critically. Even more emphatically, there is no one extractant solution which can be used to test a soil and which can always define the 'availability' of a given ion. Such extractions may be very useful guides to the supply of the ion to particular plant species in particular circumstances, but the fundamental questions concern the equilibrium concentration of the ion in the soil solutions, the chemical form in which it is present there, and the rate and degree to which this concentration is maintained near an absorbing root. In general, any change which increases the concentration of an element in the soil solution, especially that of the free ions, will increase its availability. It is obvious that such an increase in solubility also enhances the possibility of leaching and movement in the soil.

The availability of elements with variable valency (Fe, Mn, Se) differs greatly with redox potential, which depends upon soil drainage and wetness. Part of the toxic effects of anaerobic reducing soils is certainly due to the high concentrations of divalent Fe and Mn in them. However, in anaerobic soils there are also much larger amounts of organic matter in solution, which affects the solubility of all other trace metals as well. Under alkaline, well-aerated conditions Se is present as selenate, whereas in more acid and reducing conditions selenite will be produced, which sorbs on sesquioxides, and is therefore less available to plants than selenates. These relationships, and the relatively small interval between deficient and toxic concentrations in the plant, demand care in the application of Se on deficient soils.

3.7. The Root-Soil System

It is now accepted that ions move by either or both mass flow and diffusion to roots, and studies have been published on these processes in relation to uptake of Mn, Fe, B, Cu, Zn, Al (see Lindsay, 1974). This allows calculation of the necessary concentrations of the various ions in the soil solution; for example, if mass flow is to carry an adequate amount of Fe to a plant, the soil solution must be around 10^{-6} M, which is far above the equilibrium free ion concentration at normal pH values. Even with both mass flow and diffusion, the concentration needed is around 10^{-7} to 10^{-8} M, which requires the presence of chelating agents.

3.8. Practical Crop Problems in Britain

The soil factors tending to encourage deficiencies are summarised in Table II, and the practical issues are discussed in MAFF (1976; 1979) and Thornton and Webb (1980). Briefly, grassland herbage production is rarely if ever affected by trace element deficiencies - it is always the nutritional quality for livestock which is in question. The problems involved are deficiencies of Mg, P, Co, Cu, Mn, and Se, and excess of Mo. Trace element content may be affected by altering the ratios between species, in particular that between legumes and grasses, which depends upon soil and pasture management.

Boron deficiency is recognised in Britain in sugar beet and other *Beta* species, turnips, cauliflowers and lucerne. The symptoms are death or distortion of the growing points, and diminished growth of leaves and roots (MAFF, 1971; MAFF, 1976). It typically occurs on light soils, because only the water-soluble B in the soil is immediately available to the plant, and this is readily leached out. There is also

TABLE II

Factors contributing to trace element deficiencies.

Element	Factors
COPPER	Low total Cu Free $CaCO_3$ High organic matter content (e.g. peats) Low organic matter content (sands) High N, P and Zn Moisture stress High Mo and SO_4^{2-} (livestock)
IRON	High soil pH Free $CaCO_3$ High HCO_3^- High Mn Poor drainage Extreme moisture changes Liming Plant species
ZINC	Low Zn Low organic matter content Free $CaCO_3$ High pH High clay content High N, P Liming Low temperature Land levelling
MANGANESE	High soil pH Neutral-alkaline peats Free $CaCO_3$ Unconsolidated soil Poor drainage High Fe Liming Moisture stress
COBALT	Alkaline and calcareous soils High soil Fe and Mn Liming Moisture stress
MOLYBDENUM	Low soil pH High soil Fe and Al oxides High SO_4-S Leached soils Low seed Mo content
SELENIUM	Low soil pH Waterlogging and high rainfall High soil Fe oxides High SO_4-S Pasture species

After Reuter, D. J. in: Trace Elements in Soil-Plant-Animal Systems, D. J. D. Nicholas and A. R. Egan (eds.), 1975. Academic Press, London, pp. 291-324.

an antagonism between Ca and B, so that naturally calcareous soils or those recently limed, are particularly prone to the deficiency. The deficiency may be predicted, for susceptible crops, by analyses for water-soluble B, and may then readily be prevented by soil treatments with boronated fertiliser.

Manganese deficiency occurs in many crops including

cereals and sugar beet, crucifers and fruits, and on a wide variety of soil types. The major characteristics of soils producing deficiencies are that they contain large amounts of organic matter, and are of high pH, due to the complexation of Mn or its precipitation as manganese dioxide. Up to 20,000 ha receive leaf sprays with Mn, which can control the symptoms, and may give yield increases. The occurrence depends very much on local pedological conditions, and soil analysis for Mn is useless, because the amounts present are always large, and diagnosis depends upon observation of symptoms, supported by plant analyses. The soil series which are affected are well recognised (MAFF, 1976), but control of the deficiency incurs an appreciable expense. Equally, Mn toxicity is a well-recognised hazard of acid soils, though this is readily cured by liming. Cobalt is not required for higher plants, though it is essential for the Rhizobium N-fixing bacteria associated with legumes. However, no effects have been reported in Britain. The sole interest is thus in Co content in herbage. The deficiency occurs most markedly on soils derived from parent materials with low Co content, mainly in the North and West of Britain. The availability of the Co in soils is reduced by high pH, as with most of the trace metals, but sharply increased by poor drainage and wet soils, probably because of the increase in soluble organic matter, and the reduction of the Mn oxides on the surfaces of which Co is absorbed.

No Zn deficiency of any significance for crops has yet been reported in Britain, though it is of great importance in some countries. However, the toxicity due to this metal is a definite crop hazard in metalliferous zones, and similar problems arise in relation to the disposal of sewage sludge and other wastes. Extremely high contents of both total and extractable Zn may be found, though Ca is often also present in these cases and the exact cause of the plant toxicity observed may not be clear. Resistant species and lines that tend to grow on such sites are disucssed elsewhere.

Copper deficiency has been widespread in cereals in Britain, and has been extensively researched (MAFF, 1971). It occurs most frequently on calcareous and highly organic or very sandy soils, and is lessened by poor drainage or a high watertable; the reasons for this are parallel to those mentioned for Co. The deficiency prevents proper development of the grain, leading to empty ears. Soil dressings of Cu have given striking increases in yield, but these are now likely to be much less frequent, because the Cu dressing has a long residual effect, and most deficient fields have now been treated repeatedly. The Cu level in grain responds very little to soil addition of Cu (MAFF, 1971), but may affect levels in the vegetative parts of the cereal. Soil analysis has proved a useful diagnostic aid: total Cu may be a good indication when the deficiency is extreme (Thornton and Webb, 1980), and extraction with chelating solutions normally gives results which relate well to deficiency incidence. Copper toxicity is not a problem, though there are soils with high Cu levels in metalliferous areas. However Cu is undoubtedly toxic to plants when applied artificially to soils in sewage sludge or other forms.

From the annual survey of fertiliser practice carried out by ADAS it is possible to calculate approximately that the following areas received trace element supplements (Church, personal communication) Mg 100,000 ha (spray and solid); B, 50,000 ha (spray and solid); Cu 100,000 (spray); Mn, 200,000 ha (spray); Co, 17,000 ha (spray). Some areas may have multiple deficiencies, but the total area affected is clearly quite considerable.

3.9. The Availability and Need for Soil Information Relating to Geochemistry

It is obviously desirable to have information indicating the likelihood of trace element problems occurring in any area, right down to field level, based on soil measurements, and it is necessary to define which sampling and analytical methods are most dependable or relevant for any particular purpose. The methods of obtaining information are directly from clinical evidence of disease, or from field trials. For both methods the problem of the extrapolation of this information, on the basis of soils evidence, is absolutely crucial. The most extreme variations are usually related to the composition of the parent material, and this is a useful guide to their existence. However, a single parent material can vary greatly in composition and the soil-forming process can greatly modify it; thus the Lower Lias in Somerset gives rise to soils with up to 100 $\mu g\ g^{-1}$ of Mo, whereas the same formation in Glamorgan and the soils on it contain much less Mo.

Various soil analysis techniques have been used, of which the only quite unambiguous one is for total element content, but this may be of little relevance to plant growth. For this reason many workers analyse for extractable forms of the elements, though the choice of which extractant to use is largely arbitrary, and may change in the future. It has been thought unusual for total content to correlate closely with extractable values, though good correlations have recently been found for some elements within specified soil series. The ideal method is clearly to measure both total content and that extracted by some generally accepted extracting solution. For some elements (see above) a satisfactory correlation between the amounts extracted and the incidence of crop deficiency and toxicity have already been established, but it cannot be assumed that an extractant which is effective for one element will be so for others.

Using such analysis techniques, and properly based sampling schemes, detailed information is now being assembled from a few areas. The range of values is quite large even in natural soils of normal composition, and can be extreme in areas affected by old mining operations (Davies, 1980). For this reason, statistical techniques are essential in designing sampling and data handling procedures. Values based on samples from single soil pits or a few surface samples are of little use. In a detailed mathematical treatment, R. Webster (private communication) found the relationship of extractable Cu and Co contents of some Scottish soils with the series classification was rather poor, and that this depended almost wholly upon the soil parent material. The results could however be used to develop maps of the metal concentrations and of the precision of the estimated value at each point.

Further detailed sampling schemes for areas of particular interest are needed, including especially those indicated as containing exceptional levels of more than one element on the basis of stream sediment survey. The possibility of developing maps of trace element contents over large and less anomalous areas needs to be kept in mind, but the extent of variation and the required density of sampling should be closely investigated first.

3.10. References

Burridge, J. C.: 1969, 'Proc. WAAP/IBP', International Symposium on Trace Elements in Animals, Aberdeen, p. 412.

Davies, B. E.: 1980, Applied Trace Elements, John Wiley, Chichester.

H.M.S.O.: 1980, Cadmium in the Environment and its Significance to Man, Interdepartmental Report, Department of the Environment, Pollution Paper No. 17.

Johnson, H. S., Handley, J. F., and Bradshaw, A. D.: 1976, 'Revegetation of Metalliferous Fluorspar Mine Tailings', Trans. Inst. Min. Metall. A85, 32-37.

Jones, L. H. T. and Jarvis, S. C.: 1981, 'State of Heavy Metals', in: The Chemistry of Soil Processes, eds. D. J. Greenland and M. H. B. Hayes, John Wiley, Chichester, pp. 593-620.

Lag, J. and Steinnes, E.: 1978, 'Content of some Trace Elements in Barley and Wheat Grown in Norway', Meld. Norges Landbrukshøgskole, No. 10. p. 57.

Lindsay, W. L.: 1974, in E. W. Carson (ed.), The Plant Root and its Environment, University Press of Virginia, p. 506.

Loneragan, J. F.: 1975, in (eds.) D. J. D. Nicholas and A. R. Egan, Trace Elements in Soil-Plant-Animal Systems, Academic Press, p. 109.

M.A.F.F.: 1971, 'Trace Elements in Soils and Crops', Tech. Bull. 21, H.M.S.O.

M.A.F.F.: 1976, 'Trace Element Deficiencies in Crops', A.D.A.S. Advisory Paper No. 17, H.M.S.O.

M.A.F.F.: 1979, Fertiliser Recommendations for Agricultural and Horticultural Crops, H.M.S.O. (2nd edition of M.A.F.F. Bull. 209).

Mattigod, S. V., Sposito, G., and Page, A. L.: 1981, 'Factors Affecting the Solubilities of Trace Metals in Soils', in Chemistry in the Soil Environment, Amer. Soc. Agron. Madison, U.S.A. pp. 203-221.

Mitchell, R. L. and Burridge, J. C.: 1979, 'Trace Elements in Soils and Crops', Phil. Trans. R. Soc., London, B288, 15-24.

Peterson, P. J.: 1978, in (Ed.) J. O. Nriagu, Biogeochemistry of Lead in the Environment, Vol. 1B, Elsevier, Amsterdam.

Thornton, I. and Webb, J. S.: 1980, 'Regional Distribution of Trace Elements in Britain in Relation to Agriculture' in: (ed.) B. E. Davies, Applied Trace Elements. John Wiley, Chichester, p. 381.

Tinker, P. B.: 1981, 'Levels, Distribution and Chemical Forms of Trace Elements in Food Plants', in: L. Fowden, G. A. Gaston, and C. F. Mills (eds.), Metabolic and Physiological Consequences of Trace Element Deficiency in Animals and Man, Phil. Trans. R. Soc. Lond. B294, pp. 41-55.

4. GEOCHEMISTRY AND ANIMAL HEALTH

4.1. Summary

Diseases and impaired productivity due to inorganic element deficiency or excess cause significant financial loss to the British animal industry. Pathological effects and associated metabolic defects in farm livestock associated with deficiencies of Co, Cu, I, Mn, Na, P, Se, and Zn are recognised. Pathological defects may develop before overt signs of deficiency appear. Appetite, growth, resistance to infection and reproductive performance may be impaired before external signs of deficiency develop. Pathological effects due to chronic exposure of animals to the potentially toxic elements As, Cd, F, Fe, Pb, Se, and Zn are also recognised. Results of chronic low level exposure may differ markedly from those of acute intoxication. The wide range of estimated requirements of essential elements reflects both high demand at times of high metabolic activity such as rapid growth, and the effects of the composition of diet on the efficiency of absorption of individual elements from the digestive tract.

Factors modifying the response of animals to their geochemical environment comprise (i) the inorganic composition of dietary components; (ii) physiological variables including the sequestration of elements in animal tissues and the adaptation of the animal to deficiency or excess; and (iii) the effect of inorganic imbalance on the availability of elements in the diet and on tolerance of the animal population. Geochemical anomalies in soil may influence the inorganic composition of foodcrops and the proportion of different crops in the diet influences total element intake. The risk that intakes will lead to deficiencies differs between management systems based on pasture herbage and more intensive ones. Supplementary feeds rich in trace elements may obscure deficiencies that would arise from geochemical conditions.

The rate at which symptoms appear as a result of a deficient diet is influenced by the extent of tissue reserves of an element; these vary between elements and may delay the development of metabolic defects. Genetic variables influence the susceptibility of animals to deficiency or toxicity; there are marked differences between breeds in the ability to adapt to low or excessive intakes. The net requirements for many major and trace elements are significantly increased at particular times in the animals life such as at rapid growth and in late pregnancy and during lactation.

The 'availability' of elements within the digestive tract is often determined by interactions between elements of similar chemical properties or sharing common metabolic pathways. Ion absorption or translocation may be inhibited by structurally related ions; for example high intakes of Zn or Cd inhibit adsorption of Cu. Mutually antagonistic interactions between elements may promote sequestration of metals and may lead to 'secondary' deficiencies.

The most reliable means of assessing the likelihood of clinical diseases due to deficiency or excess is by the early detection of metabolic defects in the animal. The use of blood analysis to detect deficiency or toxicity does not necessarily reveal whether there are or will in due course be pathological changes. Studies of relationships between herbage composition and disease incidence have been limited in scale because of their cost. For this reason, there is growing interest in the potential animal system in studies of the distribution and origins of related diseases in animals, particularly in providing evidence of imbalance between potentially competitive elements.

Geochemical surveys can be used to identify likely problem areas providing investigations allow for the many variables that influence inorganic element flux from parent material through soil and plant to animal tissues. These variables are difficult to quantify, and this in turn complicates correlation between the distribution of geochemical anomalies and disease incidence. Nevertheless, geochemical data are useful in focussing attention in areas in which the inorganic element status of farm animals merits investigation. Preliminary geochemical data are now available for 70% of Britain. These data would seem to be of particular use in contributing to the recognition of areas in which deficiencies of Cu, Co, and Se and excesses of Mo, Se, Cu, Zn, Pb, F, and As may influence animal health and/or production. It has not yet been clearly established whether relationships exist between low Cu geochemical anomalies, low Cu in herbage and the incidence of clinical Cu deficiency in animals, though clinical Cu deficiency in cattle has been related to low concentrations of Cu in pasture herbage in some parts of northeast Scotland, and current studies in England and western Scotland suggest relationships between very low Cu geochemical anomalies and farms with a history of clinical Cu deficiency in cattle. There is however strong evidence linking high Mo anomalies delineated by geochemical mapping and both clinical and subclinical hypocuprosis in cattle. These anomalies, recognised in England, Wales and Scotland, are related to soils derived from marine black shales, lacustrine shales or Mo-rich granites.

A study initiated by the Working Party in Caithness led to a threshold value of 1 mg kg^{-1} Mo in stream sediment being recognised, above which clinical Cu deficiency was found on a high proportion of farms. Such relationships need quantitative examination in further studies, taking into account soil variables, such as drainage and pH, which influence the availability of Mo to plants. Cobalt deficiencies in cattle and sheep have been mainly found in areas where soils are derived from coarse sedimentary calcareous and acid igneous rocks. Some degree of correlation has been established with stream sediment data, though not all low Co anomalies coincide with areas where Co deficiency is found. Animal problems related to Se deficiency show marked geographical variation in Britain and would seem to be related to low concentrations of Se in herbage and cereals. However, little has yet been published on the geochemical distribution of Se in Britain, and existing evidence as to whether Se in stream sediments can predict soil or plant Se status is conflicting. Geochemical surveys have led to the recognition of high Se soils overlying marine black shales in parts of England and Wales, though chronic Se poisoning in livestock has not been identified probably because excessive accumulation of Se is only available to plants under organic high pH soil conditions such as those found in parts of Ireland. Although Zn and Mn levels in herbage and cereal crops below minimum dietary requirements for ruminants have been found on some calcareous shell-sand soils, these have not as yet been related to clinical or subclinical effects on animals.

The distribution of Pb poisoning in sheep and cattle in Britain is partly related to soils with a high Pb content. Lead ingested in soil is thought to be the main pathway to the animal. Present evidence clearly suggests that stream sediment reconnaissance data closely reflect Pb contaminated soils and are useful for the identification of areas in which risks of Pb intoxication of livestock is high. There is no evidence in Britain that consumption of herbage or soil high in Cd and Zn leads to Cd or Zn intoxication. However, geochemical survey has defined at least one area where levels of these elements in herbage are sufficiently high to induce secondary defects in the metabolism of Cu. Experimental evidence has demonstrated that both Zn and Cd are antagonists of Cu absorption and utilisation, and closer studies of the influence of high Zn/Cd geochemical anomalies on animal health are needed.

There are no instances of F deficiency in livestock recorded in Britain, and chronic fluorosis is usually caused by industrial contamination of diet or drinking water or by feeding mineral supplements rich in F. The extent to which F ingested in soil is tolerated has yet to be determined. Geochemical data could prove useful in defining areas with high F soils and with enhanced F contents of water. Geochemical surveys have focussed attention on areas with extremely high levels of As in soils in parts of Cornwall: chronic intoxication from natural sources of As has not been identified, though pathological changes are possible. Geochemical data for several other elements may also be useful for studies on animal health; these include Ca (ingestion of excess $CaCO_3$ as soil can be harmful), Fe (soluble-Fe may limit the utilisation of P and deplete liver Cu reserves and blood Cu content), Ni (assists in the metabolism of N by rumen microorganisms), and P (aphosphorosis in sheep on some low P granites).

With careful interpretation, regional geochemical data from stream sediment surveys can facilitate the identification of areas in which animals are subject to increased risk of inorganic element deficiency or toxicity. Their value can be considerably improved if factors modifying element movement from weathered rock to water or to animal diets are considered during data interpretation.

4.2. Introduction

Diseases and impaired productivity attributable to inorganic element deficiency or excess are significant causes of financial loss to the animal industry of the United Kingdom.

Incidence is not declining and economic factors necessitating increased reliance upon home produced feeds are exacerbating such problems in some areas.

There is a possibility that knowledge of the composition of the geochemical background could contribute towards the more effective identification of areas in which such losses are likely to arise. However, there are both advantages and limitations to the geochemical approach that influence the interpretation of the survey data used.

As yet, few quantitative data are available from which to assess the influence of geochemical anomalies on disease incidence and little is known about the influence of other variables upon this relationship. After outlining the nature of variables that must be taken into account when investigating the pathological response of animals to changes in inorganic element supply, this Section will consider current evidence that geochemical variability can determine susceptibility to such disorders - and will indicate situations in which further investigations appear desirable.

4.3. Inorganic Element Deficiency Diseases

The roles of essential elements are briefly summarised in Table I in which the most significant clinical consequences of deficiency are also indicated.

The following points merit particular attention when considering the influence of inorganic element supply upon disease incidence:

(i) A variety of pathological defects can develop before any clinical signs of deficiency appear, for example, connective tissue lesions and cardiac enlargement (Cu deficiency), defective leucocyte function (Se and Cu deficiencies), defective glucose utilisation (Cr deficiency).

TABLE I

Principal clinical and metabolic consequences of inorganic element deficiency diseases for which direct or indirect involvement of geochemical anomalies in aetiology is now recognised or probable.

Deficiency	Gross pathological effect	Species	Associated metabolic defect
Co	anorexia; anaemia; neurological defects	R	cobalamin-dependent enzymes (-)
Cu	defective melanogenesis; skeletal and connective tissue defects	R all	tyrosine monooxygenase (-) lysyl oxidase (-)
I	thyroid hyperplasia; reproductive failure; hair/wool loss; neurological defects	all	thyroid hormone synthesis (-)
Mo	defective keratogenesis	B	?
Mn	skeletal and cartilage defects; reproductive failure	all	chondroitin sulphate synthesis (-) ?

Table I (continued)

Na	anorexia, pica, weight loss	R, B	
P	anorexia or depressed appetite; skeletal resorption	all	
Se	myopathy (cardiac, skeletal); myoglobinuria; liver necrosis; defective leucocyte function	R C P C	oxidative damage to tissues (+)
Si	skeletal and cartilage defects	B	Si in mucopolysaccharide cross links (-)
Zn	anorexia; parakeratosis/hyperkeratosis; defective immune function	all	?

Key to species: R, ruminants; C, cattle; S, Sheep; P, pigs; B, poultry. Responses indicates thus: (-) depression of enzyme activity or of synthetic process; (+) increased concentration of product; (?) metabolic lesions not established.

Notes: (i) Deficiency of any of the above elements can inhibit growth, usually before characteristic clinical signs develop.

(ii) Deficiencies of Ca and Mg also occur in livestock but no evidence suggesting relationship of incidence to geochemical variables yet available.

(iii) Cr, V, F, Pb, and As now known to be essential for laboratory animals; evidence lacking for farm livestock as yet.

(ii) Adverse effects upon appetite and growth, upon resistance to infection or upon reproductive performance frequently arise before characteristic signs of deficiency develop. Growth failure may be the only manifestation of mild deficiency.

(iii) Although analyses of blood or accessible tissues may well reveal that the inorganic element status of an animal is low, this does not necessarily reflect the imminence or otherwise of pathological changes.

Most studies of the influence of anomalous inorganic nutrient supply upon "disease" incidence have considered either relationships to obvious clinical signs of disease or the occurrence of anomalies in blood composition. With exceptions that will be noted later, few have taken into account the incidence of 'asymptomatic' deficiency and obtained confirmation of this by assessing responses such as growth when the presumed limiting nutrient is provided.

4.4. Inorganic Element Toxicity

The principal pathological consequences arising from chronic exposure of animals to potentially toxic elements are summarised in Table II. It should be noted that the effects of chronic, relatively low level exposure that are particularly relevant in the present context, differ considerably from those induced by acute intoxication. In several instances (e.g. excesses of Cd, Zn, Cu, Fe, Ag), these include effects attributable to the induction of defects in the metabolism of structurally related ions.

TABLE II

Principal consequences of intoxication induced by chronic exposure to some elements of geochemical interest.

Element/ Compound	Effects of chronic intoxication	Suggested maximum tolerable concentration (mg kg^{-1} dry diet)*
As	incoordination; appetite failure; respiratory distress	50 (for inorganic sources) 0.2 mg l^{-1} (water)
($CaCO_3$)	poor growth; skin lesions (parakeratosis); defects in Fe, Zn, Mn absorption and metabolism	5000
Cd	poor growth (young animal exhibits enhanced susceptibility); defective wool synthesis; defects in Cu (Fe) metabolism (young animal exhibits enhanced susceptibility)	0.5 1 mg l^{-1} (water)
F	dental lesions (mottling, excessive wear); skeletal lesions (legs, ribs); poor growth and appetite	40 2 mg l^{-1} (water)
Fe	reduced growth rates; skeletal lesions; defects in Cu and P metabolism; diarrhoea	500
Pb	poor growth and appetite, young animal exhibits enhanced susceptibility; bone fractures	60
Se	lameness, hoof malformations; hair loss, growth failure	3
Zn	pancreatic and renal (young animal exhibits enhanced susceptibility) damage; anaemia; skeletal and connective tissue lesions (young animal exhibits enhanced susceptibility); perinatal mortality (young animal exhibits enhanced susceptibility)	200
Mo	secondary induction of Cu deficiency; delayed puberty and conception.	2

*) derived from National Academy of Sciences (1980) Mineral tolerance of domestic animals, Washington; (NAS): Agricultural Research Council (1980) The nutrient requirements of ruminant livestock, Farnham; Commonwealth Agric. Bureaux.

4.5. Dietary Requirements

Estimates of the minimum concentrations of individual essential elements required in the diet to maintain normal health, growth and other physiological processes are summarized in Figure 1. The wide range of estimates of requirement for many individual elements is indicated. Such variability reflects not only increases in demand associated with periods of high anabolic activity as, for example, during periods of rapid growth of the foetus or young or during lactation, but also the marked influence of dietary composition upon the efficiency with which individual elements are absorbed from the digestive tract or utilised subsequently. The influence of such variables upon pathological responses to a low trace element supply has been considered in detail elsewhere (Mills, 1981).

4.6. Factors Modifying the Response of Animals to their Geochemical Environment

4.6.1. Inorganic Composition of Dietary Components

The proportion of the total diet that is accounted for by different crops can markedly influence total inorganic element intake, the balance between individual elements and thus the probability that intake will be adequate or deficient. Thus, with respect to animal requirements, grass species are frequently low in Na, P, Mg, Cu, I, and Se and cereal grains and straws low in Ca, S, Fe, Mn, Cu, Co, and Se content. In contrast, legumes and most protein concentrates of animal or vegetable origin are either intrinsically rich sources of many essential inorganic elements or become so as a consequence of adventitious contamination during processing.

Even from such generalisations it is apparent that the

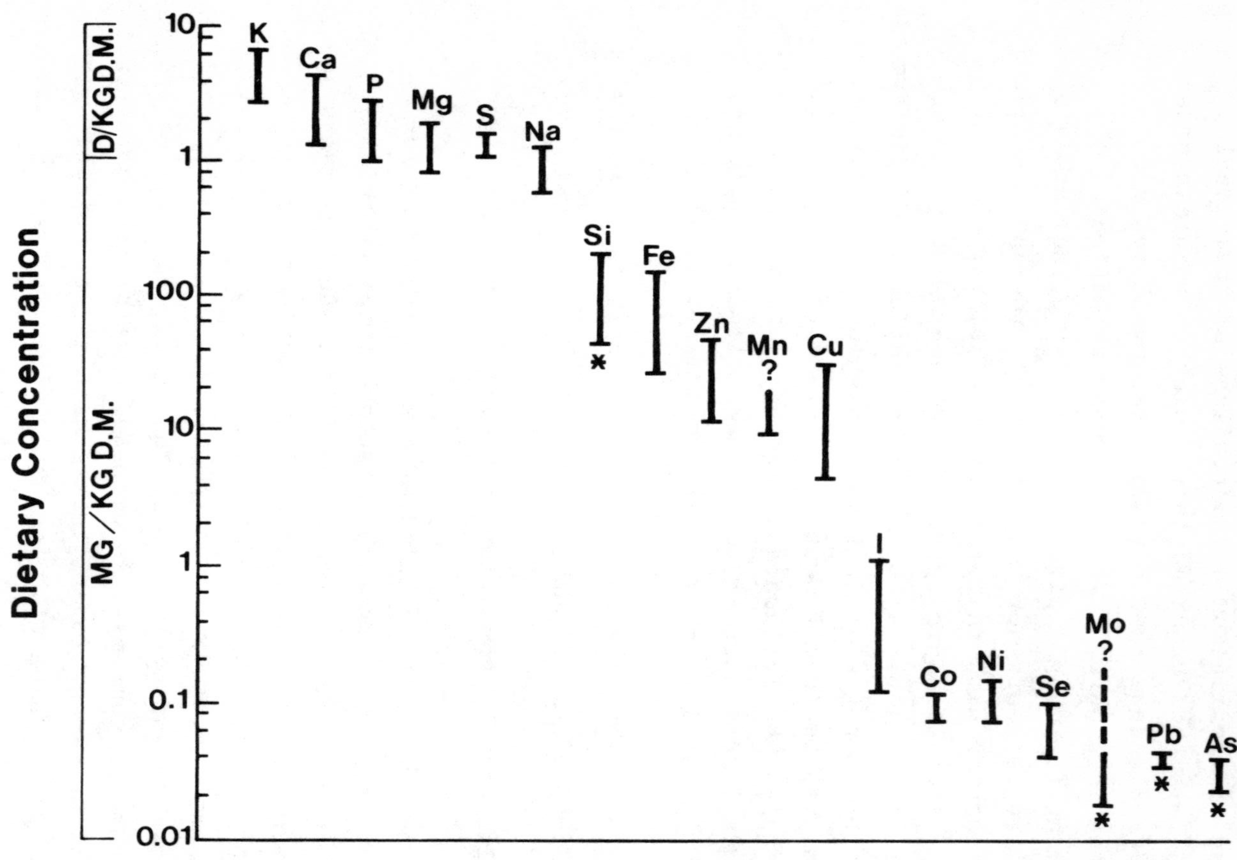

Fig. 1. Estimates of the minimum dietary concentrations of essential inorganic elements required by animals. Note: concentrations are given on a logarithmic scale. Except where indicated (*) the estimates are derived from studies with ruminant animals.

risks that intake will be deficient can differ greatly between feeding systems based primarily upon herbage and, those based upon cereals or fodder crops. It is also evident that, at common rates of useage, the supplementary feeds used by the dairy industry and other intensive systems can often obscure inorganic element deficiencies in other dietary components whether or not these originate from anomalies in the composition of soils or their parent materials.

4.6.2. Physiological variables

The rate at which metabolic defects develop and pathological changes appear once inorganic element supply becomes sub-optimal is influenced markedly by the magnitude of pre-existing tissue reserves and the rate at which these can be utilised. The rate of response to a deficient intake or to the presence of a toxic element in excess is also conditioned by adaptive changes. Once tissue reserves are depleted, such adaptation can enhance the efficiency of absorptive processes and reduce losses by excretion. Exposure to toxic concentrations of some elements in food or drinking water initiates other adaptive responses which reduce the rate of absorption of the element or restrict its access to sensitive sites within tissues. The effectiveness of such responses differs greatly between individual animal species and for individual elements. Such differences have to be taken into account when assessing the influence of anomalies in inorganic element supply upon health.

4.6.3. Sequestration of Elements in Tissues

Responses to N and K deficiencies are rapid as no tissue reserves exist. Responses to Mg and Zn deficiencies are particularly rapid when anabolic demand is high. Even though reserves of Mg and Zn exist in skeletal and soft tissues these are only utilised readily if other nutrient deficiencies or physiological stress promote increased metabolic turnover of these tissues. Substantial reserves of Fe (liver, spleen), Cu (liver), Co (as vitamin B_{12}, liver) and Se are deposited in the soft tissues of ruminants during periods of a luxus intake. The utilisation of such reserves can delay the development of metabolic defects attributable to deficiency for 2-3 months after dietary intake if any of the above elements declines below the minimal requirement. Lead and F selectively accumulate in the skeleton. Thus variables modifying the turnover of skeletal tissues influence their toxicity.

4.6.4. Partial Adaptation to Deficiency or Excess

The efficiency with which Ca, Fe and Zn are absorbed increases as deficiency develops. Losses of P and Cu are minimised as intake declines by reducing urinary excretion (P) or, in the instance of Cu, by both an increase in absorptive efficiency and reduction of biliary losses. The ability to adapt to either low or excessive intakes of Cu is subject to genetic control. Thus the susceptibility of sheep to Cu deficiency or toxicity differs markedly between breeds. Genetic variables also influence the susceptibility of cattle to Zn deficiency and of other animals to Mn deficiency. Exposure to high intakes of Cd, Zn, Hg or Cu induces the synthesis of specific proteins with a high affinity for such elements, the role of which appears to be to sequester such metals in relatively innocuous forms. This adaptive process is inhibited in Zn-deficient animals. Regulatory processes discriminating against excessive absorption of Cd, Cu, and Zn are poorly developed in young animals subsisting primarily on liquid diets. Susceptibility to Cd intoxication is particularly high if suckling animals gain access to this element.

The influence of such variables upon the response to reduced inorganic element intake is often considerable. Increases in growth rate and the demands of late pregnancy or lactation can increase between 2 and 4-fold the net requirements for many major and trace elements. Whether or not clinical manifestations of deficiency develop is conditioned not only by dietary intake but also by previous nutritional history and the facility with which the metabolic changes described above influence absorptive efficiency and the regulation of losses.

The existence of such changes in requirement and tolerance has important consequences both for the design and the interpretation of studies of the effects of anomalies in the composition of diet, soil and the geochemical environment upon health. Wider appreciation of their effects would permit more effective identification of the proportion of the animal population at greatest risk and would enhance the sensitivity of many such investigations.

4.6.5. Inorganic Element Imbalance; its Influence on Availability and Tolerance

Susceptibility to diseases attributable to inorganic element deficiency or excess is influenced not only by total intake via food and water but also by variables which influence the metabolic availability of such elements during their absorption and redistribution within tissues. In many instances, availability is determined by competitive interactions between elements exhibiting similar chemical properties or sharing common metabolic steps within the animal. Data on the inorganic composition of the diet or of soils and their parent materials can often suggest the involvement of such interactions in the aetiology of disease and frequently influence decisions as to the most effective means of control.

4.6.6. Reactions Influencing Solubility within the Digestive Tract

Although the quantity of an element ingested is an important determinant of its concentration with the digestive tract, the fraction that is present in aqueous or lipid phases and thus is potentially available for absorption, is influenced by the processes illustrated in Table III. Those particularly relevant to the interpretation of geochemical data are indicated (*).

Co-precipitation reactions initiated by the microbiological generation of sulphide from dietary sources of sulphur within the rumen of herbivores influence the metabolic fates of several elements, particularly Cu. Much of this S^{2-} is normally absorbed rapidly and reoxidised. However, recent evidence suggests that the strong inhibitory effect of Mo on Cu absorption arises from intrarumenal sequestration of reactive sulphide as Mo/S complexes and their later reaction with Cu to yield insoluble, unabsorbable. products. This role of Mo may not be unique as the potency of high dietary Fe as a Cu antagonist also appears to be contingent upon S intake. Furthermore, *in vitro* studies now indicate that a range of insoluble metallic sulphides can undergo substitution reactions with soluble Cu at physiological pH (Mills, 1984) and, like Mo and probably Fe, sequester S in forms that potentially react with Cu and convert it to physiologically unavailable forms. Such phenomena may well account for the hitherto inexplicably high incidence of Cu deficiency disorders in many areas adjacent to metalliferous mineral deposits and their workings.

Evidence of the adverse effect of high phosphate intakes upon utilisation of Fe, Mn, Ca, and Zn is confined

TABLE III

Processes influencing inorganic element solubility/availability from the gastrointestinal tract.

Enhancing efficiency of utilisation	Depressing efficiency of utilisation		
recycling in salivary and other secretions	Mutual precipitation reactions		
Na, K, I, P, Zn	Initiated by:	Precipitated element	
	(high SO_4^{2-})* high S^{2-}	Cu, Pb	**
Synthesis of biologically active forms by microorganisms	high CO_3^{2-}	Mg, (Zn?) (Mn?)	**
	high Fe	P (Cu?)	**
Co	high PO_4^{3-}	Fe	
	Organic ligands (e.g. phytate)	Zn, Mn, Cu, Fe	-
Release from insoluble ligands by H^+, amino and hydroxyacids			
Mg, Cu, Fe, Mn, Zn, (Ca?)			

Notes: Reactions initiated by changes in the composition of the ingesta that directly or indirectly can reflect anomalies in the composition of soils or the geochemical background are indicated **. Detailes are given in the text.

Evidence of a direct influence of clay minerals (e.g. clinoptinolite) upon solubility/availability is conflicting and thus has been excluded.

*Effects on metal utilisation initiated by S^{2-} can develop in ruminants when SO_4^{2-} is ingested and reduced subsequently by rumen microorganisms.

to situations in which diets contain >20 g p kg^{-1}. Thus, it is unlikely that the contribution of P from geochemical sources would be sufficient to induce such effects.

Several studies have indicated that dietary concentrations of $CaCO_3$ in excess of 30 g kg^{-1} can adversely affect growth and induce a range of metabolic defects in cattle, sheep and pigs.

4.6.7. Competitive Interactions Influencing Tissue Distribution, Function or Toxicity

Mechanisms of inorganic element absorption particularly relevant to the development of deficiency on decreased tolerance of potenitally toxic elements have been discussed by Mills (1980). Whether absorption or translocation of an ion occurs by energy dependent pumping against a concentration gradient or by carriers involved in metal/ ligand-exchange reactions, many such processes are subject to competitive inhibition from structurally related ions. Typical examples are the inhibitory action of SO_4^{2-} on the absorption of MoO_4^{2-} and, conversely of MoO_4^{2-}, SeO_4^{2-} or AsO_4^{2-} on SO_4^{2-} activation and transport.

The presence in tissues of potentially toxic concentrations of Cd, Zn, Cu, or Hg induces the synthesis of SH-rich proteins (metallothioneins) with a high but relatively non-specific affinity for the above group of metals. This initiates a series of mutually antagonistic interactions between these elements. Thus, induction of metallothionein synthesis by an excess of Zn or Cd promotes not only the sequestration of these metals but may sufficiently inhibit the utilisation of Cu to initiate metabolic defects attributable to a secondary deficiency of Cu.

Also relevant are interactions reflecting the metabolic interdependence of pairs of metallic ions. The influx and efflux of Fe from intestinal mucosa, liver and many other tissues is mediated by enzymes in which Cu has an essential functional role. Hence, the risks of development of defects in the metabolism of Fe are strongly enhanced by Cu deficiency. Absorption and tissue retention of Cd and Pb and thus susceptibility to toxicity increases markedly in animals even marginally deficient in Fe or Ca; the metabolic defects responsible for these effects have not been identified.

The influence of many such interactions upon the availability of essential elements is reflected by the relatively wide range of many estimates of minimal requirements indicated in Figure 1. The extent of their effect upon susceptibility to deficiency disease and on the tolerance of many toxic elements is indicated in Table IV which has been compiled solely from evidence indicating that the interaction described has metabolic or pathological significance.

Such data indicate the particular value of multielement analysis in studies of the incidence and origins of related diseases in animals. Evidence of the existence of anomalies in the balance between potentially competitive elements can often suggest appropriate lines for further enquiry in cases where clinical symptoms are insufficiently specific to permit unequivocal diagnosis. In instances where the antagonist is not absorbed and its locus of action is primarily confined to the digestive tract as, for example, the induction of Cu deficiency by high environmental concentrations of Mo, the origin of consequent pathological changes can rarely be resolved by analysis of animal tissues alone. Their aetiology may only become apparent after the detection of imbalances in the composition of the diet or, after allowing for the modifying influence of soil/plant relationships upon such imbalances, from the detection of anomalies in the inorganic composition of soils and their parent materials.

TABLE IV

Metabolic interactions modifying susceptibility to inorganic element deficiency or chronic toxicity in animals.

DEFICIENCY SYNDROME	INORGANIC DIETARY COMPONENT PROVOKING OR EXACERBATING SYNDROME a)
Co	$++CaCO_3$*
Cu	+Mo*, +S*, +Fe*, +Cd*, +Zn*, +Ag*
I	+Co*, (+As), (+F)
Fe	$++PO_4^{3-}$*, +Cu*, +Zn*, +Cd* -Cu*
Mn	+Ca, $+PO_4^{3-}$
Mg	++K*
Mo	++W
P	++Fe*+
Se	+S (in analogous chemical species)*, +Cu*,+Ag*
Zn	++Ca*, +Cu*, +P (as phytate)
TOXICITY SYNDROME	
Cd	-Fe, -Cu
Cu	-Fe*, -Zn*, (-Se)
F	(+Mo or -Cu?)
Mo	-Cu*, +S (ruminants)*, $-SO_4$ (non ruminants)
Pb	-Ca*, -Fe, -S (ruminants), (-Se)
Zn	-Cu, -Fe

a) Key:

(i) + or ++, <u>increased</u> dietary concentrations of component exacerbate biochemical or pathological manifestations of syndrome.
(iii) -, low concentrations of component exacerbate.
(iii) observations awaiting confirmation are indicated in parentheses ( ).
(iv) *interaction reported to be significant under field conditions.

4.7. Prediction of the Risks of Deficiency or Toxicity from Geochemical Survey Data

It is recognised by the Working Party that the early detection of pathologically relevant metabolic defects in a representative population of animals will always provide the most reliable basis from which to assess the likelihood that clinical disease attributable to inorganic element deficiency or excess will arise. However, such an ideal approach demands investigative resources that are rarely available and usually preclude its use for the purposes of survey work. Many instances of inorganic element deficiency or toxicity have been detected from evidence of related changes in the inorganic composition of blood. However, from such evidence alone it is rarely possible to determine whether pathological changes will develop eventually or already exist. Such uncertainty often arises in the interpretation of survey data for blood Cu, Se, Fe, Pb, Cd and Mo, and this difficulty will remain until more is known of the factors which influence relationships between the inorganic composition of blood and that of tissues sensitive to deficiency or excess (Mills, 1981).

Until such problems are overcome, the principal objective of investigational work must be to identify those aspects of environment and management that enhance risks of deficiency or excess and thus justify prophylactic measures. Evidence considered earlier suggests that the ability to anticipate and eliminate many such risks would be enhanced if local or regional anomalies in the inorganic element composition of herbage or other staple feeds for animals were identified. Although such data are being accumulated by some Advisory Service Laboratories, resources are limited, progress is slow and surveys are small in scale. No such surveys are being undertaken even in many countries suffering far greater economic losses from such diseases than the United Kingdom. The extent to which data derived from geochemical surveys can be used to indentify likely problem areas depends on the ability of investigators to recognise and allow for the effect of the many variables that influence inorganic element flux from parent material through soil and plant to animal tissues. Because it may often be difficult to quantify the effect of such variables, it is unrealistic to expect that high orders of correlation between the distribution of geochemical anomalies and that of disease incidence will normally exist. Since the primary value of geochemical data will be to focus attention upon areas within which the inorganic element status of the animal population merits closer investigation using more direct and sensitive techniques, this limitation may often be acceptable.

General information on element distribution is available for 70% of the British Isles. With extension and refinement, such data offer the possibility of defining not only areas with inherently low or high background concentrations of essential or toxic elements but also those in which inorganic element imbalance and antagonism may play a role in the aetiology of morbidity or disease in animals.

4.8. The Influence of Geochemical Anomalies upon Animal Health: an Appraisal of Field Evidence

4.8.1. Copper Deficiency

Whether relationships exist between the distribution of low Cu geochemical anomalies, low Cu in herbage or crops and the incidence of clinical Cu deficiency in animals has not yet been studied adequately. A subjective appraisal of relationships between the distribution of beef cattle herds with low blood Cu (< 0.7 mg Cu l^{-1}) and that of Cu in stream sediments in Northern Ireland has been made by Thomson and Todd (1976). Low blood Cu was rarely encountered were stream sediment Cu exceeded 56 mg kg^{-1}; most low-Cu herds were located in areas where sediment Cu was < 28 mg kg^{-1} although some normal herds were also found in such areas.

Clinical Cu deficiency in cattle attributable to extremely low concentrations of copper (< 3 mg Cu kg^{-1} DM) in pasture herbage occurs in some areas of North East Scotland where the total Cu content of soils derived from quartzites and some granites is abnormally low. A statistically significant relationship between the incidence of low Cu geochemical anomalies and that of low blood Cu in cattle has emerged in recent studies conducted by Thornton et al. (to be published). Geochemical data for some coastal margins adjoining Western Scottish sea lochs (I.G.S., to be published) suggests an association between very low Cu anomalies and farms with a past history of clinical Cu deficiency in cattle. Statistical evaluation of this presumed relationship is impracticable since most of such farms have ceased rearing cattle because of nutritional problems.

Evidence of the involvment of high-Mo geochemical anomalies in the aetiology of concurrent Cu deficiency and acute molybdenosis in ruminants first arose from studies in areas with calcareous soils derived from high Mo shales of the Lower Lias. Clinical signs of Cu deficiency in cattle are frequently encountered in other areas where soils are derived from a variety of argillaceous sediments or marine black shales relatively high in Mo content but differing widely in geological age (Thornton and Webb, 1980).

Thornton et al. (1972a) investigated the relationship between high Mo stream sediment anomalies (> 3 mg Mo kg^{-1}) in areas overlying Namurian or Visean marine shales in North Staffordshire and the frequency with which low blood Cu (< 0.7 mg Cu l^{-1}) was detectable in cattle herds of this and adjacent low Mo areas. Of 350 animals investigated, 77% grazing pastures within the high Mo area had low blood Cu compared with only 37% in normal Mo control areas. A particularly notable finding from later studies (Thornton et al., 1972b) was that marked increases in growth rate (10–70%) resulted when Cu was given to cattle on each of 6 farms within areas associated with high Mo anomalies. Clinical signs of Cu deficiency had not been evident on any of these farms and thus the possibility that a Mo-induced deficiency of Cu was limiting the growth rate of cattle had not been considered previously.

A further indication of the value of data on stream sediment Mo content for identification of localities within which clinical Cu deficiency may arise in cattle was obtained during a study initiated by this Working Party. Retrospective data on the distribution of farms within Caithness with a previous history of copper deficiency were obtained both from practising veterinarians and from the Regional Veterinary Services Laboratory, Thurso. This was compared with data for the geochemical distribution of Mo derived from stream sediment analyses (I.G.S., 1979) and subsequently smoothed by a moving average surface proximation technique (NERC/IGS, G-EXEC Programme). This study covering 76 normal farms and 66 with a record of Cu deficiency indicated that selection of an appropriate threshold value for definition of a molybdenum-anomaly influenced both the discriminatory power of the survey and proportion of Cu-deficient herds located. The most satisfactory compromise was achieved using a Mo threshold of 1 mg kg^{-1} in stream sediments. The ratio of deficient to identified normal herds was 4.9 times higher within areas with a high Mo anomaly than outwith such areas. Eighty two percent of farms with a record of clinical Cu deficiency were correctly identified using this criterion.

In this ad hoc study, as in those described previously, no weighting was applied to geochemical data for Mo to reflect its increased significance in the aetiology of Cu deficiency in ruminants when high soil moisture, low soil redox potential and relatively high soil pH enhanced uptake of this element by herbage (Plant and Moore, 1979; Thornton and Webb, 1980). Adequate consideration of such variables would undoubtedly enhance the sensitivity of such an approach. Even so, the veterinarians participating in the Caithness study regarded its outcome as quite adequate to indicate situations in which animals were at risk and should thus be examined for covert or overt signs of Cu deficiency.

4.8.2. Cobalt Deficiency

The influence of soil type upon the incidence and severity of pining (i.e. Co deficiency) in sheep was first described in 1807. Areas of the United Kingdom with a high incidence of Co deficiency in cattle and sheep frequently have soils derived from Old Red Sandstone, Triassic sandstones, granite or rhyolites. Incidence varies on soils derived from calcareous parent materials or from siliceous materials of Ordovician or Silurian origin, such variability presumably reflecting both variations in Co content and the extent to which locally high concentrations of Ca, Fe, and Mn inhibit Co uptake by herbage (Section 3).
From a geochemical survey of Co distribution in Counties Wicklow and Carlow (Eire) it was concluded that Co deficiency in sheep was most prevalent on strongly leached granitic soils formed in situ and particularly where stream sediment Co was < 7 mg kg^{-1}. Incidence was moderate or sporadic when stream sediment Co was 7-15 mg kg^{-1} or when glacial over-burden or peat covered low Co granites. A similar association between low Co anomalies (< 10 mg Co kg^{-1})

in granitic areas of Devon and Cornwall and the incidence of Co deficiency in sheep was described by Thornton and Webb (1970). Kiely et al. (1969, 1978) found a significant relationship between the Co content of soils and stream sediments in two studies in Eire and indicated subsequently that the incidence of Co deficiency in sheep is particularly high in areas where low Co anomalies exist in sediments derived from Carboniferous limestone.

It is evident however that not all low Co anomalies coincide with areas in which Co deficiency occurs in ruminants. The evidence that a high soil moisture content and low pH both enhance Co uptake by plants even when total soil Co is low may well be relevant to this situation. It has also been pointed out that local coprecipitation of Co and Mn may occur in stream sediments in areas where the availability of soil Co is low and clinical Co deficiency exists in animals, and there is additional evidence of the occasional involvement of high Mn anomalies in the aetiology of Co deficiency.

In some areas, the use of acid extraction procedures to determine available Co in soils has provided a satisfactory basis from which to assess the likely risks of Co deficiency. The extent to which low available Co in soils is predictable from geochemical data and knowledge of relevant soil conditions (e.g. pH, moisture content) has not been assessed adequately. Both from the evidence presented above and from an ad hoc study of data from 145 farms in the Highland Region on behalf of the Working Party it appears probable that the incidence of Co deficiency may well bear some relationship to the distribution of low Co geochemical anomalies. However, sufficient quantitative data are not yet available to assess the significance of this relationship nor to determine the extent to which the precision of such a study might be improved by concurrent consideration of other variables known to influence the flow of Co from weathering rock to soil, plant and animal (Plant and Moore, 1979).

4.8.3. Selenium Deficiency

The incidence of skeletal and cardiac muscle myopathy attributable to selenium deficiency shows marked geographical variations within the United Kingdom. Surveys of the blood Se content of ruminants show similar geographical variability. Present evidence indicates that the incidence both of clinical disease and of the less readily characterised failure of growth now known to be a further consequence of Se deficiency reflects the frequency with which Se concentrations of fresh or conserved herbage and cereal grains and straws are low (< 0.02 mg Se kg^{-1} dry matter). No statistical appraisal of the validity of such presumed relationships has yet been made nor has it been determined why the Se content of crops in some areas is so low.

A survey conducted in Scotland indicated that statistically significant growth responses obtained by the administration of Se to sheep were confined to localities with soils derived from granitic or arenaceous parent materials. However, more recent investigation of the incidence of low blood Se in cattle herds in North East Scotland (Arthur et al., 1979) failed to reveal any clear association between low Se status and the character of soil parent materials as defined by reports of the Scottish Soil Survey.

In view of the economic importance of Se deficiency in some localities it is regrettable that so little is known of the geochemical distribution of Se within the United Kingdom. Evidence as to whether surveys of the Se content of stream sediments will provide an adequate basis for the prediction of soil or crop Se content is conflicting (e.g. Webb and Atkinson, 1965; Kiely et al., 1968). The availability of soil Se to crops is influenced by soil redox potential and is related inversely to the soil content of SO_4^{2-} and Fe as indicated in Section 3 but the significance of these variables as determinants of crop Se content under practical agricultural conditions is largely unknown.

4.8.4. Selenium Toxicity

Although geochemical surveys have led to the recognition of seleniferous soils overlying some marine black shales in Central and South West England and Wales, instances of chronic Se intoxication in livestock have not yet been identified within the United Kingdom (Thornton and Webb, 1980). Studies in Eire of the incidence of selenosis in cattle and horses indicate that the extent of bio-concentrations within lacustrine depostis of Se derived from Se-rich parent materials and the subsequent oxidation of their selenides or organo-Se fractions enhance both the uptake of Se by crops and the incidence of selenosis in cattle and horses. Webb and Atkinson (1965) have pointed out that geochemical evidence of the distribution of high Se anomalies is useful for the identification of high risk areas provided the influence of such variables upon Se accumulation by crops is taken into account.

4.8.5. Zinc and Manganese Deficiencies

Herbages and cereal straws with Zn and Mn contents substantially lower than the minimum dietary concentrations normally required by growing ruminants (approximately 15 mg Zn or Mn kg^{-1} D.M.) have been found in several recent surveys. Most were obtained from areas with calcareous shell-sand (machair) soils. As yet, there are no indications that consumption of such feeds provokes clinical manifestations of Zn or Mn deficiencies. No studies have yet been made of their possible effects upon growth and skeletal structure, both of which are known to be influenced by such deficiencies before clinical symptoms appear.

4.8.6. Lead Toxicity

The geographical distribution of chronic Pb poisoning of grazing livestock with the United Kingdom is closely related to that of soils with a high Pb content. The Pb content of herbage in areas overlying high Pb soils is markedly enhanced during winter senescence but little higher than normal during summer growth. Recent evidence suggests however that Pb from ingested soil can account for up to 80% of the total Pb intake of grazing stock and that bovine Pb closely reflects soil Pb content (Thornton and Kinniburgh, 1978). Additional evidence that the Pb content of stream sediments closely reflects that of soils in contiguous areas clearly suggests that this geochemical survey technique would be appropriate for the identification of areas within which the risks of Pb intoxication of livestock are inherently high.

Although the incidence of chronic Pb intoxication in livestock in the United Kingdom is relatively low, it is evident that farming practices in many high Pb areas have been adapted to minimise the risks of intoxication either by reducing the period of exposure of young stock or, in some areas, by complete withdrawal of livestock from such areas. Better definition of such areas by more detailed soil or stream sediment surveys, and the use of supplementary feeds rich in Ca, P, and S to inhibit lead absorption (Table IV) during the brief periods of high sensitivity to intoxication would markedly enhance their potential for livestock production.

4.8.7. Cadmium/Zinc Toxicity

There are no convincing indications that consumption of herbage or inadvertent ingestion of soils in the proximity of high Cd/high Zn anomalies within the United Kingdom sufficiently influences the retention of Cd or Zn by livestock to provoke pathological changes directly attributable to Cd or Zn intoxication.

It is evident, however, that the Cd and Zn contents of herbage in at least one such area defined by geochemical surveys are sufficiently high to induce secondary defects in the metabolism of Cu (Table IV). Whether a high recorded incidence of clinical Cu deficiency in ruminants from areas adjacent to or overlying mineralised deposits in the Mendips or the sphalerite-rich marine black shales of Derbyshire and North West England is thus attributable to high Cd/high Zn geochemical anomalies in these areas has not yet been investigated. It is clear, however, that the incidence of Cu deficiency in these areas is not caused by a low Cu intake nor in the instance of the Mendips and North Derbyshire, by the existence of high Mo anomalies inducing a Mo/Cu antagonism.

There is clear experimental evidence that Zn and, particularly Cd, are potent antagonists of Cu absorption and utilisation and both elements have been incriminated in the aetiology of Cu deficiency arising in areas subject to industrial contamination. A closer study of the influence of high Zn/high Cd geochemical anomalies upon animal health is warrantable.

4.8.8. Fluorine

Although there is now clear evidence of the essentiality of F, no instances of F deficiency in livestock have yet been recorded. Chronic fluorosis of livestock is usually caused by contamination of the diet with F-rich dusts of industrial origin, by use of mineral supplements rich in F or by consumption of drinking water containing more than 3 mg F l^{-1}. The extent to which F derived from ingested soil is tolerated has not been determined. Soluble F in drinking water that has permeated strata relatively rich in fluorapatite is poorly tolerated. Identification from geochemical survey data of localities with enhanced F contents in water accessible to livestock or used for irrigation would provide a useful indication of the need to exclude other sources of F from the diet. There is a particular need for such information in many developing countries in which the incidence of fluorosis in livestock is high and its effects sometimes exacerbated in localities where other geochemical anomalies exist.

4.8.9. Arsenic Toxicity

High concentrations of As (23-2080 mg kg^{-1}) occur in some soils of the Tamar Valley, Cornwall. In some, As concentrations exceed the threshold value (1000 mg kg^{-1} dry soil) found in localities in New Zealand in which chronic As poisoning occurs in ruminants. The incidence of As intoxication in such areas has been attributed either to ingestion of soil As or access of animals to drinking water containing > 0.2 mg As l^{-1}. Other studies have shown that susceptibility to As intoxication is high when soils or stream deposits contain arsenic limonite but substantially less if adventitious As_2S_3 or As_2S_2 was ingested. Conditions promoting a low redox potential in soils enhanced incidence.

Although chronic As intoxication from natural sources of As is unknown in the United Kingdom, the possibility that high As anomalies of the Tamar Valley and other areas could induce covert pathological changes cannot be excluded.

4.8.10. Other Inorganic Elements

The calcium ion is relatively well tolerated by animals. However, excessive consumption of $CaCO_3$ affects growth rate adversely and can precipitate secondary deficiencies of P, Fe, and Mn (and, in non-ruminants, Zn deficiency) if the dietary intake of these element is marginal with respect to normal requirements. The suggested maximum tolerable intake of $CaCO_3$ (5% of diet) by most species of farm livestock could frequently be exceeded if adventitious contamination of herbage with calcareous soils exceeded approximately 10% of the diet. Soil can account for up to 40% of the dry matter intake of sheep grazing during winter and up to 10% for cattle. The influence of this on the utilisation of other essential elements by animals grazing in calcareous areas has not been investigated adequately. The value of geological, geochemical and soil surveys for identification of areas of interest is self evident.

Evidence from studies with ruminants and monogastric animals indicates that excessive ingestion of Fe in forms that become soluble during digestion in the rumen or stomach can sufficiently limit the utilisation of P to inhibit growth and induce skeletal defects. Ingestion by cattle of soluble or potentially soluble sources of Fe, whether these are derived from pasture, its contaminating soil, from irrigation water or from experimental supplementation of the diet is now known to induce rapid depletion of liver Cu reserves and blood Cu content if Fe intake exceeds 500 mg kg^{-1} diet. Interest in the role of Fe in the aetiology of clinical Cu deficiency is growing rapidly and variables influencing its biopotency as an inhibitor of Cu utilisation are being investigated. The extent to which geochemical and soil composition data can contribute to field and experimental studies of this topic will depend upon broad definition of the species of Fe compound present in high Fe anomalies.

Evidence of the essentiality of Ni now extends to recognition of its role in the metabolism of N sources by rumen microorganisms and thus in the N economy of ruminants. Deficiency thus aggrevates the effect of a low N intake upon growth. It is highly probable that investigation of the significance of Ni deficiency in livestock will be initiated in the near future. Whatever unrecognised limitations may influence the value of geochemical data for identification of low Ni areas for such initial studies, no alternative data exist.

Many variables other than the total P content of soils or their parent materials influence the availability of P to plants and thus supply to herbivores. Thus data describing the geochemical distribution of P will normally be of little value for the identification of areas in which the risks of P deficiency are high in livestock. There are, however, strong suspicions that the incidence of aphosphorosis in sheep in areas with strongly acidic soils derived from some granites may be related to the low P content of such parent materials. Geochemical data for P distribution might well contribute to the more effective definition of such areas.

4.9. Conclusions

A frequently complex array of variables can modify the

influence of the geochemical environment on animal health and productivity. Despite this, there are some indications that geochemical investigations can provide data that are of value for assessing the risks that anomalies in inorganic element supply may either induce clinical disease or adversely affect the inorganic element status of animal populations. While the geochemical approach to the assessment of risk is relatively imprecise - a characteristic frequently shared by approaches such as soil or dietary compositional surveys - this limitation must be balanced against relatively low costs and its yield of information describing, for the long term, the primary chemical characteristics of the soil forming elements from which the principal inorganic element supply is derived.

While more systematic and quantitative appraisals of the potential of geochemical data for this purpose are undoubtedly required it is also emphasised that future studies should take greater account of the effects of other variables known to influence the susceptibility of animals to changes in inorganic element supply. More appropriate consideration of such modifying factors during the interpretation of "raw" geochemical data could greatly enhance the value of this promising approach.

4.10. References

Arthur, J. R., Price, J., and Mills, C. F.: 1979, 'Observations on the Selenium Status of Cattle in N.E. Scotland', Vet. Rec. 104, 340-341.

I.G.S.: 1979, Regional Geochemical Atlas: South Orkney and Caithness, London: Inst. Geol. Sci.

Kiely, P. V. and Fleming, G. A.: 1969, 'Geochemical Survey of Ireland; Meath-Dublin Area', Proc. Roy. Irish Acad. 68B, 1-28.

Mills, C. F.: 1980, 'Trace Elements in Animals', in: S. H. U. Bowie and J. S. Webb (eds.), Environmental Geochemistry and Health, London: the Royal Society, p. 51.

Mills, C. F.: 1981, 'Some Outstanding Problems in the Detection of Trace Element Deficiency Diseases', Phil. Trans. R. Soc. B, 294, 199.

Mills, C. F.: 1984, 'The Influence of Chemical Speciation on the Absorption and Physiological Effect of Trace Elements from the Diet or Environment', Proc. Dahlem Konferenzen (in press).

Plant, J. and Moore, P. J.: 1979, 'Regional Geochemical Mapping and Interpretation in Britain', Phil. Trans. R. Soc. B. 288, 95-112.

Thomson, R. H. and Todd, J. R.: 1976, 'The Copper Status of Beef Cattle in Northern Ireland', Brit. J. Nutr. 36, 299-303.

Thornton, I., Kershaw, G.F., and Davies, M. K.: 1972a, 'An Investigation into Cu Deficiency in Cattle in the Southern Pennines, I. Identification of Suspect Areas Using Geochemical Reconnaissance'. J. Agric. Sci. Camb. 78, 157-164.

Thornton, I., Kershaw, G. F., and Davies, M. K.: 1972b, 'An Investigation into Copper Deficiency in Cattle in the Southern Pennines, II. Response to Copper Supplementation', J. agric. Sci. Camb. 78, 165-171.

Thornton, I. and Kinniburgh, D. G.: 1978, 'Intake of Pb, Cu, and Zn by Cattle from Soil and Pasture', in: M. Kirchgessner (ed.), Trace Element Metabolism in Man and Animals - 3, p. 499.

Thornton, I. and Webb, J. S.: 1970, 'Geochemical Reconnaissance and the Detection of Trace Elements Disorders in Animals', in C. F. Mills (ed.): Trace Element Metabolism in Animals - 1, Edinburgh; Livingstone, pp. 397-407.

Thornton, I. and Webb. J. S.: 1980, 'Regional Distribution of Trace Element Problems in Great Britain', in B. E. Davies (ed.): Applied Soil Trace Elements, London, Wiley, pp. 381-439.

Webb, J. and Atkinson, W. J.: 1965, 'Regional Geochemical Reconnaissance Applied to some Agricultural Problems in Co. Limerick, Eire', Nature, Lond. 208, 1056.

4.11. Appendix: Relationships between the Distribution of high-Mo Geochemical Anomalies and of Cu Deficiency in Cattle in Caithness, N.E. Sutherland

The role of dietary Mo in the aetiology of Cu deficiency of ruminants and that of variables modifying its effects have been described in Sections 2-4. The following *ad hoc* investigation was undertaken to determine whether, if the effects of such variables were ignored, quantitative relationships were still evident between the distribution of high Mo geochemical anomalies and that of Cu deficiency in cattle.

An area of Caithness/N.E. Sutherland for which records of the past incidence of Cu deficiency were available was selected. Veterinary practitioners indicated on O.S. maps the locations of (i) farms for which there was unequivocal evidence of the need for Cu prophylaxis and (ii) farms for which there was no evidence of Cu deficiency. Sixty-six farms were thus identified. Corroborative evidence was obtained from records of the Regional Veterinary Service Laboratory, Thurso.

Data for the geochemical distribution of Mo derived from stream sediment analyses (I.G.S., 1979) were smoothed using NERC/IGS Programme G-EXEC and plotted as a grey-scale map using 0.1 km^2 unit cells. This, and the identified locations of farms with Cu-deficient and normal cattle herds are presented in Figure 1.

The data describing the distribution of Cu-deficient and normal herds do not cover the entire population of such animals. The extent to which the sub-populations so classified accurately reflect the proportions of deficient and normal animals over the entire area indicated in Figure 1 is not yet known. Analysis of data was thus confined to examination of the influence of arbitrarily defined Mo threshold values upon (i) the ratio of $\frac{\text{Cu-deficient}}{\text{normal}}$ herds identified in high or low-Mo locations, (ii) assessment of the overall discriminatory power of each Mo threshold and (iii) the proportion of the total number of locations with a history of Cu deficiency found in areas with a high Mo geochemical anomaly. Results are presented in Tables I and II. These data suggest that information on the distribution of high Mo geochemical anomalies provides a satisfactory basis from which to anticipate the involvement of Mo in the aetiology of Cu deficiency of cattle in the above area and adequately indicates localities in which the risks of development of this disease are high. The discrimination achieved by threshold values of either 1 or 3 mg Mo/kg^{-1} is regarded as adequate to justify direct investigation of the Cu status of animals located in areas where Mo concentrations exceed either of these thresholds. The data from this investigation are inadequate to determine either the optimum threshold Mo concentration for this area or the extent to which this may differ locally through the influence of variables described elsewhere in this Report. Such aspects merit investigation.

FIGURE 1.

Stream sediment Mo/Cu deficiency in Caithness cattle

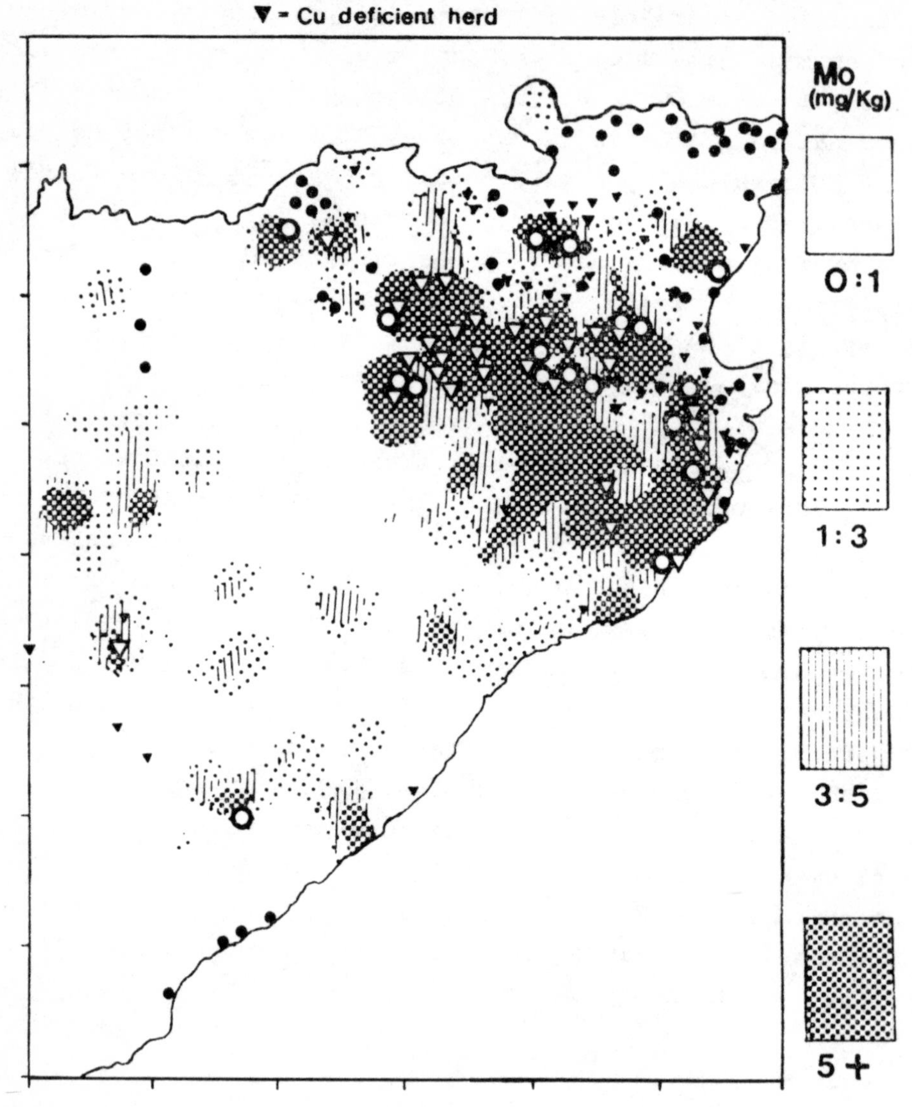

TABLE I

Relationships between the distribution of high Mo geochemical anomalies and that of farms on which clinical Cu deficiency is recorded in cattle (Caithness, N.E. Sutherland).

Definition of Mo-Anomaly (mg kg^{-1} stream sediment)	Distribution of Cu-Deficient (66) and Normal (76) Farms				% of total Cu-deficient farms located within anomaly
	Within high Mo anomaly		Outwith high Mo anomaly		
	Deficient	Normal	Deficient	Normal	
> 1 Mo	54	36	12	40	82
> 3 Mo	44	23	22	53	67
> 5 Mo	32	18	34	58	48

(χ^2 for Mo-thresholds 1, 3, and 5 mg kg^{-1}; 18, 16, and 9 respectively $P < 0.01$).

TABLE II, Discriminatory power of geochemical data for Mo for assessing distribution of Mo-induced Cu deficiency in cattle

Definition of Mo-Anomaly (mg kg^{-1} stream sediment)	No. of Cu-Deficient Herds per 100 Normal Herds Located		Discrimination Ratio
	Within high Mo anomaly (a)	Outwith high Mo anomaly (b)	(a/b)
> 1 Mo	172	35	4.9:1
> 3 Mo	222	47	4.7:1
> 5 Mo	188	58	3.2:1

5. GEOCHEMISTRY AND HUMAN HEALTH

5.1. Summary

The health of man may be affected by the amounts of chemical elements available from food, drinking water and the atmosphere. In developed societies, the relationship between local geochemistry and human intakes of these chemicals in food and water may be of very low order due to complex systems of food distribution and water supply: the contrary may be the case in developing countries were local communities live closer to the land. The complexity of interrelationships between both major and trace elements and health poses a further problem. Care must be taken not to attribute causal significance to correlations or associations between diseases and geochemical factors until this is conclusively proven.

The relationship between I deficiency and endemic goitre has been firmly established in many countries, though other environmental factors are also of importance and have yet to be thoroughly investigated. The relationship between low F intake and increased dental caries is a good example of the direct effect of natural geochemistry on human health, though other elements, including Se, Mo, and Pb, may be of importance in the protective process. Where water is high in F, mottling of teeth and skeletal fluorosis may occur.

Cardiovascular diseases are a major cause of death in adults in most developed societies and are related to dietary factors, cigarette smoking and hypertension. Incidence and mortality vary both between different countries and between different regions in a country, and have been linked with the geochemical environment. The Regional Heart Study jointly funded by the Medical Research Council and the Department of the Environment is examining the geographic variations in cardiovascular morbidity and mortality in Britain with reference to environmental factors including the quality of drinking water. Water hardness, rainfall, temperature and socio-economic factors have been shown to be significant. Cardiovascular mortality was 10-15% higher in areas with very soft water than in those with medium hard water, though a clear mechanism of causality has yet to be identified. In Britain water hardness is correlated with Ca, CO_3 and Si: soft acid waters from upland areas are corrosive and may carry trace metals in solution. Clinical studies are continuing in 24 towns with the aim of relating cardiovascular events to both personel risk and environmental factors. Calcium makes a major contribution to water hardness and water could be an important source of Ca to man. Magnesium is also present in some hard waters which may contribute a significant proportion of many daily intakes. In Britain the association between Mg in drinking water and cardiovascular mortality is insignificant, probably because water hardness in Britain is primarily related to Ca content rather than that of Mg. However, in Canada and the United

States, Mg-rich water is relatively common and a relationship has been suggested. There is no known relationship between cardiovascular mortality and water Na levels. Clearly, some factor closely associated with water hardness plays a small but significant role in cardiovascular mortality.

There are marked geographic variations in the mortality and incidence patterns of some cancers, and it is considered that geochemical factors might be causally implicated. It has been suggested that high nitrate ingestion may be linked with human cancer, as nitrites derived from nitrates may in vivo form N-nitroso compounds which are potent carcinogens. However, a comprehensive study in Britain has shown no relationship between nitrate levels in drinking water and mortality from cancer. Soil organic matter content and the ratio of Zn:Cu in soil has been empirically related to gastric cancer. Further studies on relationships between cancer and geochemistry are unlikely to provide clearcut answers, but may uncover contributory factors in the development of some cancers.

Relationships between multiple sclerosis and geochemistry have been suggested and the elements Cu, Mo, and Pb possibly implicated. Possible links between trace metal imbalance and neurological problems are worthy of further study.

For some time it has been considered that areas of Britain defined geochemically as anomalously high or low in one or more elements could be in some way related to human health. However populations in such regions are often sparse and mortality data insufficient to be used as an index. Nevertheless, they provide populations in which trace element intake and effects on blood and tissue levels can be assessed and physiological measurements made. The old zinc mining area of Shipham provides a good example of a geochemical anomaly, where garden soils and housedusts are considerably enriched in Cd, Pb, and Zn. Although some vegetables have elevated levels of Cd and Pb, no obvious health effects have been found; it is emphasised however that this study has not yet been completed. The elements Cr, Co, Cu, F, I, Fe, Mn, Mo, Ni, Se, Si, Sn, V, and Zn are essential components of the human diet; non-essential elements such as Pb and Cd, present in food and water, may be toxic.

Many trace elements reported to be associated with human disease do not meet criteria necessary for establishing causal relationships. Lead is elevated in many soft drinking waters and is thus a possible factor in cardiovascular diseases; in Britain evidence linking water Pb, blood Pb and hypertension/renal disorders is not conclusive. A variety of studies into the possible effect of Pb on intellectual development in children show contradictory results. The main sources of Cd to man are food and cigarette smoking; water plays a minor role. Possible long-term health effects of low level intakes of Cd have been extensively studied but results are inconclusive; Cd levels are raised in the kidneys of hypertensives and there are possible interactions with Zn. Although is is considered that the Zn intakes of many Americans is marginal, there is no evidence in Britain of a relationship between Zn intake and health, though a protective role in cardiovascular disease has been suggested. Chromium is primarily obtained from food sources, is necessary for glucose and lipid metabolism, and has been implicated as a protective factor in heart disease and atherosclerosis, and as well as a cause of digestive tract cancers as a result of high-level occupational exposure. There is no evidence that Cr is a health hazard in non-occupational exposure. Selenium is an essential nutrient and intake is related to diet. Evidence from China relates Keshan disease, a heart muscle disease in children and women, to low (deficient) intakes of Se. Associations of gastro-intestinal cancers and breast cancers with Se intake are controversial.

There is insufficient information to causally link levels of trace elements in blood with health and disease. Abundant experimental evidence points to numerous analytical errors and erroneous values as a result of contamination. Normal levels of many trace elements in blood have not yet been established.

5.2. Introduction

The health of animals and man are known to be affected by the amounts and properties of the chemical elements available from food, drinking water and the atmosphere. In developed societies, however, deficiencies or excesses of elements in a population group are difficult to relate to the direct geochemical environment because of our well-developed systems of food distribution and our far-ranging water supply systems. Indeed the relationship between local geochemistry patterns and the local chemical intakes in food and water may be of a very low order. A further problem in the unravelling of the possible relationship between environmental geochemistry and human disease lies in the complexity of the inter-relationships and interactions between the various bulk and trace elements. Simple unifactorial studies focussed on one specific element, to the disregard of related or inter-acting elements, are destined to produce inadequate or misleading information.

Another major problem arising from unifactorial studies, is that of attributing causal importance to significant associations. 'Association is not causation' and a number of criteria should be satisfied before correlation or association is accepted as having causal significance (Hill, 1965). While strong support for a causal relationship may come from experiments, contrived or natural, such interventions in human populations and their interpretation require considerable knowledge of the natural history of the human disorder as well as careful factorial design and analysis.

There are some literature reports which demonstrate that the natural occurrence of trace elements can directly affect human health. Arsenic in drinking water has caused As poisoning , gout has been explained by dietary Mo, Se poisoning has been reported and there is one recorded case of supposedly simple Co deficiency in an infant girl.

We are at present confronted with the fact that the major causes of death in our society - cardiovascular diseases and the cancers - do not have firmly established causes and have very uncertain control. There is a growing concensus that naturally occurring trace elements might influence human health through the agencies of air, water and food. This belief is reinforced by the discovery of the essential nature of a number of elements for animal health (e.g. chromium, selenium) and by the better understanding of the role of the essential metals in metabolic processes. In addition, there are a few well-established examples of human health being affected by the amounts and properties of the chemical elements available from the environment. Iodine and F have long been known to influence health directly through food or drinking water although an examination of these relationships shows that even these situations are not as clear-cut as we often assume.

5.3. Iodine and Fluorine

5.3.1. Iodine

The relationship between endemic goitre and deficiency of iodine in soil and water has been firmly established in many countries. Together with F, it is perhaps one of the best pieces of evidence for a direct link between environmental geochemistry and human health. However, a high goitre prevalence in areas with appreciable quantities of iodine in the water and differing prevalences of goitre in areas with similar iodine water levels, suggest that other factors may also be involved in the aetiology of this condition. Not only are there natural anti-thyroid substances (goitrogens) in certain vegetables but their goitrogenic potential may vary according to where they are grown. There have also been suggestions that other elements (F, As, Co) may interact with I in the production of goitre but this situation has never been fully explored.

5.3.2. Fluorine

Dental epidemiology provides convincing evidence that trace elements in the environment can affect the health of whole communities and presents a good example of the direct effect of natural geochemistry on human health. Once F is incorporated into teeth it reduces the solubility of the enamel to acidic material and provides significant protection against dental caries. The situation appears to be sufficiently clear for the fluoridisation of water supplies to be actively encouraged in many countries, and yet it is of interest to note that other elements have also been implicated in the protective process. In the United States, increased intakes of Se have been associated with higher rates of dental caries and high levels of Mo have been associated with reduced rates of dental caries. Studies in

Britain suggest that dental caries is more prevalent in areas where the soils have a high Pb content and this could be of considerable importance for studies of the effect of Pb on children, based on collections of teeth and analysis of their lead content. Fluorine, Se, Mo, Pb, Se, and V can all enter the crystal lattice of hydroxyapatite (of which dental enamel is an example) and alter its physical properties. It might be well not to regard the trace element-dental caries relationship as having been solved as there is considerable possibility that elements other than F may be of importance.

Adverse effects may arise in areas where the water is naturally high in F and these relate to mottling of the teeth and skeletal fluorosis. For nearly thirty years studies have been carried out to assess whether there is any link between F in the water and cancer. The fear of such a relationship has been considered in particular by those opposed to fluoridation of drinking water supplies. At present, the concensus is that there is no acceptable evidence that F in water is carcinogenic to human beings.

5.4. Cardiovascular Disease

Cardiovascular diseases (coronary heart disease, hypertension and stroke) constitute the major cause of death in adult life in most economically advanced countries of the world. International comparative studies suggest that dietary factors, cigarette smoking and hypertension account for much of the striking variation between countries. There are also marked variations in the incidence and mortality rates of cardiovascular disease between the different regions within a country, e.g. United States, Great Britain. Studies of such international differences have linked cardiovascular disease with the geochemical environment, specifically through the medium of the drinking water supply (Masironi 1979). Studies in Japan 25 years ago drew attention to a close relationship between stroke deaths and the acidity (sulphate/carbonate ratio) of river waters which was in turn related to the geochemistry of the catchment area. Since the Japanese report, many other studies have reported a negative association between water hardness and mortality rates from cardiovascular disease; areas supplied with soft waters tend to have higher rates of cardiovascular disease than those supplied with hard waters. The negative association is clearly seen in studies covering wide geographical areas with a wide range of water hardness and involving large numbers of people. More limited studies, which compare two or three towns, or several counties within a province, have produced conflicting results, usually revealing no association between water hardness and cardiovascular disease. Emerging from the many epidemiological studies is the suggestion that all the main types of cardiovascular disease are involved in the water story. Much of the criticism and uncertainty concerning the water story relates to the lack of a clear mechanism whereby water quality (hardness) affects cardiovascular mortality. Whether hard water is protective (i.e. associated with protective factors) or whether soft water is injurious is uncertain, and whether the bulk elements (Ca, Mg) or the trace elements are central factors, has not yet been determined. This lack of specific knowledge should not militate against consideration of water quality playing a role in human health, although it may urge caution in attempts to change the water quality in a specific manner.

The Regional Heart Study (Pocock et al., 1980) is examining the geographic variations in cardiovascular mortality in Great Britain with special reference to the possible effects of environmental factors, including the quality of drinking water. In a study of cardiovascular mortality (1969-1973) in 253 towns in England, Wales, and Scotland, five factors were identified which substantially explained (in statistical terms) the geographic variations in cardiovascular mortality. These factors were water hardness, rainfall, temperature and two socio-economic factors (percentage of manual workers and car ownership). After adjustment for climatic and socio-economic factors, cardiovascular mortality in areas with very soft water (around 25 mg l^{-1}), was estimated to be 10-15% higher than in areas with medium-hard water (around 170 mg l^{-1}), while any further increase in hardness was not associated with any further reduction in cardiovascular mortality. Water hardness affects both stroke and coronary heart disease mortality but has no effect on non-cardiovascular mortality e.g. cancers. Thus, there appears to be a 'water factor' related to cardiovascular disease and one that is closely related to water hardness.

In Great Britain, water hardness correlates closely and positively with a number of other water variables such as Ca, NO_3, and Si and also (negatively) with the percentage of water derived from upland sources. These waters tend to be soft and acid and in them, metals are mostly bound to humic material and to hydrous manganic-ferric oxides. If it is soft water that is injurious to cardiovascular health, then the effect of humic substances or the corrosive effects on plumbing systems with subsequent carriage of trace metals, could be of critical importance.

The further phases of the Regional Heart Study should elucidate these relationships further. They include a clinical survey of some 8000 middle-aged men in 24 towns in England, Wales, and Scotland and their follow-up over a 5-10 year period in order to relate cardiovascular events to both personal risk factors (blood pressure, smoking, etc.) and to environmental factors (water quality, etc.).

5.5. Calcium, Magnesium, and Sodium

Bulk elements in water, derived in the main from geochemical sources, which have been invoked in explaining the relationship between water quality and cardiovascular disease are Ca and Mg. Calcium and Mg usually account for most of what is called water hardness but the contribution made to total hardness by these two elements may vary considerably from place to place. Sodium in drinking water usually makes a very small contribution to total Na intake but recent work has focussed attention on possible effects on blood pressure in young people.

5.5.1. Calcium

Calcium is an essential element for man. Less than 30% of the Ca ingested in solid food is absorbed and thus water could be of importance as a source of Ca. In 21 major European cities, the proportion of total Ca provided by drinking water averaged 17% (2% to 28%). The evidence for water Ca being of major consequence for human health is mainly derived from the negative association between water hardness and cardiovascular disease and the major contribution made to total water hardness by Ca. It is possible that the importance of water Ca is a negative one in that hard waters are not as corrosive as soft waters and are less likely to leach out potentially toxic elements such as Pb, Cd, and Cu. It has been demonstrated that the concentration of Ca in a medium inhibits absorption of Pb, Cd, Zn, and Cr and thus the concentration of these elements relative to Ca may be more important than the absolute values of either.

It is possible that for certain population groups living on Ca-poor diets, the Ca in drinking water could provide a critical addition to total Ca intake. Conversely, the coincidence of Ca-poor diets with low-Ca drinking water could be critical. It is often stated that the weakest link in the relationship between water Ca and cardiovascular mortality is the lack of a plausible biological mechanism, but there seem to be no lack of hypotheses in this field. Some of these are specifically related to Ca, others to complex inter-relationships between Ca and other elements, e.g. Mg, Na.

5.5.2. Magnesium

Magnesium is an essential element in all body tissues and is intimately concerned with the function and structure of the myocardium. In experimental animals Mg deficiency can induce death of heart muscle tissue. The average dietary intake of Mg in western countries is reported to be at or below the recommended total intake. There is some evidence that the daily Mg requirements are not readily met in many industrial societies, and if this is so, then water Mg could contribute a useful, if small, amount. In Canada, drinking water from areas with a high Mg concentration, may contribute up to 20% of the total daily intake, compared with around 1% in water areas with a low Mg concentration. In Europe, the average concentration of Mg in drinking waters is relatively low and contributes about 10% of the daily intake.

The hypothesis which relates Mg deficiency to high risk of cardiovascular disease has strong support in animal experiments but is not strongly supported by epidemiological evidence. Studies in Canada and the United States have claimed that Mg relates more strongly to cardiovascular mortality than does Ca. In the British Regional Heart Study, the association between Mg in drinking water and cardiovascular mortality is insignificant. The explanation for these inconsistent findings probably lies in basic geochemical differences between the geographic regions. In the United States, hardness of drinking water is just as closely associated with Mg levels as with Ca levels. In Great Britain, there is a poor correlation between water Ca and water Mg concentrations and the levels of total water hardness are primarily determined by the water Ca concentration. This lack of consistency would suggest that Mg is not the critical factor in the 'water story' and that we should continue to look at other factors closely related to total hardness.

5.5.3. Sodium

Few sources of drinking water in Great Britain are naturally high in Na. However, the common methods of softening drinking water, ion-exchange or soda-lime processes, substitute Na for Ca, Mg, and other ions. The daily intake of Na varies tremendously between societies, some living apparently well on less than 1g a day, other consuming 10-12 g day^{-1}. It is unlikely that the amount of Na derived from drinking water makes any appreciable difference to total intakes in most societies (probably 2-6% in Europe) and it is uncertain as to whether Na in drinking water is absorbed more readily than Na in solid food.

The studies of Tuthill and Calabrese (1979) in the United States, have related differences in the mean blood pressure levels of high-school children to differences in the Na content of the drinking water. While these studies have generated considerable interest, examination of the data does not allow a firm conclusion to be drawn, as the role of dietary Na itself has not been satisfactorily considered. In the British Regional Heart Study, no relationship is evident between cardiovascular mortality and water Na levels on a town comparison basis, nor is there any relationship seen between water Na levels in a town and the mean blood pressure of middle-aged men in those towns.

5.6. Cancers and Geochemistry

As with cardiovascular disease, there are marked geographic variations in the mortality and incidence patterns for specific cancers. It has been suggested that much of this variation is due to environmental factors capable of inducing or influencing tumour formation, and some of these factors may be inorganic substances occurring in the natural environment. Many of the early studies in Britain relating cancers to geological or geochemical phenomena failed to take into account the specificity of different cancers or the detailed characteristics of the populations being studied. Nevertheless, the distribution of site-specific cancers has been well-documented in Britain and these maps have stimulated some research into geochemical factors which might be causally implicated. Studies of soil conditions and cancer incidence have suggested that soils with significantly higher organic content might be linked with gastric cancer and the ratio of Zn to Cu in the soil has also been implicated in the gastric cancer story. While it is true that relationships between the cancers and environmental geochemistry remain largely unexplored in Great Britain, it would be equally true to say that studies in this field are unlikely to produce clear answers. It is not expected that we will identify a series of elements responsible for specific cancers; it is possible that we might uncover some contributory factors in the development of some cancers and this could be of considerable importance. However, even such rewards will not be easy to achieve.

5.7. Multiple Sclerosis and Geochemistry

The case for suggesting that trace elements in the environment might affect the occurrence of diseases of the nervous system, especially multiple sclerosis, is founded partly on the observation that certain heavy minerals can affect nervous tissue. Current views on MS suggest that it is the result of a viral infection in temperate climates, but that the virus lies dormant until it is triggered or until there is some failure of an immune mechanism. The disease bears some resemblance to swayback in lambs, where Cu deficiency in the pregnant ewe is known to underlie the cause. It seems that the primary lesion in swayback is a low content of Cu in the brain, leading to a deficiency of cytochrome oxidase in the motor neurones. There are also suggestions that Pb toxicity might be a factor in MS in Britain and analyses of garden soils and vegetables from areas of different prevalence rates of MS have shown raised levels of Cu, Zn, and Pb. Blood lead studies have however not supported this hypothesis. The fact that the Cu deficiency in swayback can be induced by a high Mo intake has led to proposals that high risk MS areas are those where Mo is retained in the soil against leaching and is more available for plant uptake. The whole area of neurological and neuro-muscular problems in humans and possible links with trace metal imbalances seems worthy of further study.

5.8. Anomalous Geochemical Areas

Surveys of metal contents in stream sediments (especially work at Imperial College, London and at the IGS) and other less extensive surveys of soil metal contents (U.C.W., Aberystwyth) have provided local and regional maps for part of Great Britain. These maps have revealed certain localities where one or more trace elements occur at levels significantly in excess of average levels for the whole country. These anomalous areas may occur because of unusual bedrock conditions or because of environmental contamination arising from the mining and smelting of ore deposits. Epidemiological studies in several of these anomalous areas are providing useful information on the problems inherent in using such sites to substantiate the postulated link between geochemistry and human health.

The areas which are being, or have recently been studied in some detail include north east and west Wales, north Somerset, the Tamar Valley and the Peak District of Derbyshire. The populations of those areas are relatively small; most of the inhabitants live in small towns and villages and the remainder comprise a thinly distributed rural population. In consequence, it is difficult to standardise either morbidity or mortality data for age and their use as a possible index of health is thereby limited. For the most part these areas draw their water from relatively distant reservoirs rather than local aquifers. They tend to be upland in nature so that vegetable growing is confined to private gardens whilst vegetables offered for sale are derived, together with other foodstuffs, from distant sources. Consequently, the dietary link with local geochemical conditions is weak. On the other hand, these villages have often developed from earlier mining settlements and, in several instances, new housing development has occupied reclaimed mine waste areas.

Two anomalous areas are presently being investigated. Stream sediment work and soil analyses have revealed high concentrations of Cd and Pb in agricultural and garden soils in Shipham, Somerset and in Halkyn and certain other nearby villages in Clwyd. In both localities drinking water is derived from outside the area and metal concentrations appear normal. But levels of Cd and Pb are considerably elevated in some vegetables, but not all. In Shipham, household dust contains above normal levels of these metals and a dust survey is being made in Halkyn. No health effects are evident in any of the long-term residents of these villages which can be attributed to Cd or Pb but studies of blood and urine are in progress (Inskip et al. 1982). These studies are providing useful information on the problems inherent in using anomalous sites to unravel the complexities of trace metalhuman health relationships.

The National Food Survey has provided information on household expenditure and has indicated the value of garden or allotment fruit and vegetables as a percentage of total expenditure on fruit and vegetables. In rural areas (0.5 persons acre^{-1}, or 0.2 persons ha^{-1}) 35% of the value of fruit and vegetables is grown on allotments or gardens compared with 8% in urban areas (7 persons acre^{-1} or 3 persons ha^{-1}) and, overall, fruit and vegetables account for 17% of the total expenditure on food. OPCS census data for 1971 show that 5% of the total population live at a density of 0.2 ha^{-1}. In the standard regions of Great Britain the proportion of fruit and vegetables grown on allotments/gardens is 7% in Wales and the North and North-West England, 12-15% in Scotland, York-Humber, the Midlands, Anglia and the South-East and 22% in South-West England.

Despite the difficulties and limitations outlined above it seems evident that rural populations in geochemically anomalous areas might provide information on trace element intake and of element levels in blood or urine or of various physiological parameters related to body trace element burdens. Dental epidemiological investigations in Wales, Somerset and the Tamar Valley have revealed a consistent association between dental caries and raised levels of Pb in soil and food but not with Cd. In Derbyshire a significant relationship has been observed between Pb in blood and hair of children exposed to soils ranging from 420-14000 mg Pb kg^{-1} and in Ceredigion raised blood Pb in children has been ascribed to inhalation or ingestion of fine mine wastes containing 14,000 mg Pb kg^{-1}. In contrast, downwind of a deflating tailings area at Y Fan mine no effect was observed on blood Pb levels of adults. In Shipham, Somerset, garden soils contain up to 100 mg Cd kg^{-1} and 2600 mg Pb kg^{-1} and Cd and Pb concentrations are also raised in vegetables and household dusts. No health effects are evident in the long-term residents which can be attributed to environmental Cd or Pb. Similarly, in Clwyd, garden soils, vegetables and dusts in Halkyn and nearby villages are contaminated by Cd and Pb and the results of studies of blood and urine are expected to be available in 1982.

5.9. Trace Elements and Health

In a remarkably clear review of the problems inherent in trying to establish relationships in man between geochemistry, trace elements and human health, Underwood (1979) discusses the reasons for these difficulties, and in particular, the criteria which must be satisfied before association is accepted as causation. For many of the trace elements for which associations with disease have been reported, the evidence does not meet the exacting criteria necessary for establishing causal relationships (Hill, 1965), but it does stimulate further research into the geochemical environment in relation to human disease. It is not possible to review every trace element and its reported associations with disease but we draw attention to a few specific elements which are currently attracting interest and attention.

5.9.1. Lead

Lead occurs naturally in the Earth's surface, although it is sometimes difficult to distinguish between soils that have acquired Pb only from natural sources and soils that are polluted by mans' activities. Drinking water may contain Pb derived from the soils through which the water has travelled and in addition, may acquire Pb from the distribution system. Under some circumstances, the concentration of lead in drinking water can become extremely high. Lead pipes in plumbing and Pb-lined storage tanks can play a major determining role, but acidity, hardness and temperature of water all have to be taken into account. The volume of water flushed through the system before sampling is also critical.

The major reason for the pursuit of Pb in drinking water as a possible factor in cardiovascular disease is the presence of elevated Pb levels in many soft drinking waters and the association of soft waters with higher cardiovascular disease rates. Studies in Glasgow have related high blood pressure to increased blood lead concentrations in males and the same workers have implicated water lead levels to dimished renal function in elderly subjects. The evidence linking water Pb, blood Pb and hypertension/renal disorder from these two studies is open to criticism (Shaper, 1979) but the high prevalence of plumbosolvent waters in Great Britain and their association with higher cardiovascular mortality rates makes the continued search for better evidence of considerable importance.

Some studies have suggested that Pb derived from various sources, including water, may affect intellectual development in children. A wide variety of studies have been carried out with contradictory results and the unsatisfactory methodology in many of these studies makes it difficult to draw firm conclusions. The studies of Needleman and his colleagues (1979) in the United States have in particular occasioned considerable interest and controversy. This paper and others have been critically received in the DHSS report on "Lead and Health" which concludes "Together these studies provide some evidence of an association between raised tooth dentine Pb levels and a slight lowering of measured intelligence" (DHSS 1980). As the Needleman study draws its conclusions from 158 children selected in various complex ways from an original sample of 3329 children, the DHSS view is probably too charitable. However, the issue remains a real one and because of the relatively high intake of Pb, in many parts of Great Britain, particularly in Scotland, this issue must be investigated further.

5.9.2. Cadmium

Cadmium is a relatively rare element and is usually found in association with Zn; the average concentrations in the Earth's surface are approximately 0.1 and 60 ppm respectively. The weathering of rock and soil probably contributes only a small proportion of the Cd found in water supplies, relative to the amounts derived from sewage effluents, industrial waste and the decomposition of airborne particulates. Water put into public supply receives purification treatments which tend to decrease Cd concentrations, but several of the materials used in piping may be sources of additional Cd contamination. There is little evidence that corrosion contributes much Cd to the final drinking water and although Cd concentrations at tap may be higher than in the original water supply, they nevertheless tend to be very low. In adults, food and cigarette smoking are the important sources of Cd intake; water plays only a very minor role. In perspective, a person smoking 20 cigarettes a day will absorb 20-40 times as much

Cd from cigarettes as he would obtain from drinking water.

There is very little information on the possible long-term health effects of low level intakes of Cd, although it is probably the most extensively studied metal in relation to cardiovascular disease. Animal studies (rats, dogs) have repeatedly shown that extremely small amounts of Cd produce a rise in blood pressure and that Zn or chelating agents will lower the blood pressure. In animals, the ratio of Cd to Zn in the kidney appears to be more important than the concentration of Cd itself. In man, Cd is higher in the kidneys of hypertensives than in normotensives, but this could be the result of previous treatment for hypertension. A review of all the data on Cd in drinking water concludes that they are too fragmentary to permit an evaluation of its possible role in cardiovascular disease. It seems unlikely that Cd derived from geochemical sources plays a role in human health.

5.9.3. Zinc

Zinc is a ubiquitous element and although there appears to be a strong relationship to river water hardness in the United States this does not appear to hold for tap water. It seems that both geochemistry and plumbing corrosion play a role in determining tap water concentrations of Zn, and that their relative contributions will vary considerably from place to place. A review of the available data seems to favour a protective role in cardiovascular disease for Zn, but it must be emphasised that the data are of very mixed quality and the correlations reported have usually been insignificant.

There is some concern in the United States that Zn nutrition of many Americans is marginal, particularly for certain groups on low intakes of animal protein and those using a large amount of processed foods. Zinc deficiency in crops is widely recognised in the United States but not in Britain and there is no evidence in Britain of Zn intakes being associated with any specific aspect of ill health.

5.9.4. Chromium

Most rocks and soils contain small amounts of Cr, usually in a highly soluble form. Much of the more soluble Cr in soils is the result of contamination by industrial emissions. The levels of Cr found in water are generally low and higher levels tend to be associated with waters of greatest hardness. Drinking water can be a significant source of Cr in some areas, but most Cr is obtained from food sources (meat, fish, vegetables).

Chromium appears to be necessary for glucose and lipid metabolism and for the utilisation of amino acids in several systems. It has also been implicated as a protective factor in atheroscelerosis and diabetes mellitus. As hard water tends to be associated with higher Cr levels, the association between hardness and cardiovascular mortality has been used to support the view that Cr protects against heart disease. Direct associations between Cr in drinking water and cardiovascular death rates are weak and inconsistent. In high doses in occupationally exposed individuals, Cr has been implicated as a cause of digestive tract cancers in man. The threshold level of exposure to Cr which will produce health effects is not clear and there is no evidence that non-occupational exposure to Cr constitutes a health hazard.

5.9.5. Selenium

There is considerable variation in the levels of Se in soil

and in the vegetation in different parts of the world, due to varying geochemical conditions. Intakes of Se are mainly conditioned by dietary patterns and only to a limited extent by Se in drinking water. Selenium is an essential nutrient in several animal species and certain endemic diseases of farm animals have been identified in low Se areas which have been effectively prevented by Se supplementation. There is increasing evidence that Se may be essential for humans, and studies in China suggest that Keshan disease (a heart muscle disease in children and women) may be related in part to low intakes of Se. High Se intakes have also been implicated in human disease but the evidence is not convincing. Epidemiological studies from the United States suggest that mortality from certain gastro-intestinal cancers tends to be higher in low Se areas and that there is a some association with breast cancer in women. These controversial correlations should be treated with caution (Underwood, 1980) but they do indicate the need for such suggestions to be investigated further.

5.10. Blood Levels of Trace Elements

During the past twenty years, there has been considerable interest centred on the plasma or serum levels of trace elements in health and disease. Progress in analytical methodology has provided sensititve techniques for trace element measurement and yet the availability of information reveals a situation which is far from satisfactory. A recent review examines published information on the levels of 18 trace elements in human blood plasma or serum of apparently healthy individuals (Versieck and Cornelis, 1980). There are serious inconsistencies in the reported concentrations of trace elements, even when similar methods are used, and disparities of several orders of magnitude are not uncommon. Although some differences may reflect true biological variations, abundant experimental evidence points to analytical errors. Specimen contamination is responsible for a considerable number of erroneous values and much of the problem can be laid at the door of sampling and sample handling. "The most careful analysis of a contaminated sample remains a futile exercise that cannot but worsen the existing confusion". This survey shows with alarming clarity that we are far from being able to establish the normal levels of most trace elements.

Almost all the studies described in this extensive review (265 references) relate to very small numbers of subjects and in none of these studies are the blood levels of trace elements used in a prospective attempt to relate the initial levels to the subsequent manifestation of disease. As this is probably the only convincing method of determining what is biologically normal (i.e. conducive to good health or an absence of disease) as opposed to statistically normal, one can only regret the present state of information.

References

DHSS: 1980, 'Working Party on Lead in the Environment', Lead and Health, HMSO, London.

Hill, A. B.: 1965, 'The Environment and Disease: association or causation?', Proc. R. Soc. Med. 58, 295-300.

Inskip, H., Beral, V., and McDowell, M.: 1982, 'Mortality of Shipham Residents Forty Year Follow-up', Lancet, 17 April 1982, 896-889.

Masironi, R.: 1979, 'Geochemistry and Cardiovascular Diseases', Phil. Trans. R. Soc. Lond. B288, 193-203.

Needleman, H. L., Gunnoe, C., and Leviton, A. et al.: 1979, 'Physiological Performance of Children with Elevated Lead Levels', N. Eng. J. Med. 300, 689-695.

Pocock, S. J., Shaper, A. G., Cook, D. G. et al.: 1980, 'British Regional Heart Study: Geographic Variations in Cardiovascular Mortality, and the Role of Water Quality', British Medical Journal 280, 1243-1249.

Shaper, A. G.: 1979, 'Cardiovascular Disease and Trace Metals', Proc. R. Soc. Lond. B205, 135-143.

Tuthill, R. W. and Calabrese, E. J.: 1979, 'Elevated Sodium Levels in the Public Drinking Water as a Contributor to Elevated Blood Pressure Levels in the Community", Archives of Environmental Health 34, 197-203.

Underwood, E. J.: 1979, 'Trace Elements and Health: An Overview', Phil. Trans. R. Soc. Lond. B288, 5-14.

U.S. National Committee for Geochemistry: 1979, 'Geochemistry of Water in Relation to Cardiovascular Diseases', Washington DC. National Academy of Sciences.

Versieck, J. and Cornelis, R.: 1980, 'Normal Levels of Trace Elements in Human Blood Plasma and Serum', Analytica Clinica Act. 116, 217-254.

6. CONCLUSIONS

Over the past twenty years progress in geochemical research has led to a better understanding of the distribution of elements in rocks and soils and of the processes that control their redistribution in the surface environment. However, little systematic information is available on the geographic distribution of elements in soils, plants and waters, such as would be necessary for spatial relationships between geochemistry and health to be established. In particular, data on levels of trace element abundances in these media are considered to be essential.

Information on trace element distribution has been met in part by geochemical reconnaissance surveys undertaken by Imperial College and the regional geochemical maps of the Institute of Geological Sciences. Geochemical atlases published for England and Wales, Northern Ireland and parts of Scotland show the distribution of up to 37 elements in stream sediments, and these reflect the contents in bedrock, overburden and soil. Most of this new geochemical information is available in computer-readable form and, with careful interpretation by geochemists, provides a useful data base to relate to information on plant, animal and human health.

The major element requirements of plants are largely known and corrective measures can be taken to eliminate deficiencies. Trace element requirements are less well understood. Their availability to plants is related both to total concentration and to chemical and physical soil characteristics such as pH, and the chemical speciation in the soil solution, and to plant factors such as growth rate and genetic effects.

The physical chemistry of most trace elements in the

soil is poorly understood, as is the formation and behaviour of organo-metallic complexes and the detailed mechanisms at the root-soil interface.

The geochemistry of soil parent material can be a useful indicator of the likelihood of trace element problems occurring in an area, and regional geochemical surveys can highlight areas for immediate attention as they denote the total levels of trace elements in soils, particularly in areas with strong geochemical relief. However, direct sampling of soils is necessary to obtain detailed information.

The aetiology of diseases attributable to inorganic element deficiency or excess in livestock is often complex. For this reason it is unrealistic to expect that high risk situations can be recognised effectively if investigative action is confined to the definition of the inorganic element status of animal tissues, to the analysis of diets or soils or to the study of the composition of the geochemical environment. All these individual approaches have limitations which reduce their applicabilty. Many reveal anomalies in the supply of elements that could have relevance to disease incidence but none, used in isolation, provide sufficient information to determine the origins of the problem, its likely geographical distribution and its ultimate pathological and economic significance.

Data on the geochemical background can often make a contribution to such investigations. For example, evidence is growing that regional geochemical data can identify areas in which animals are subject to increased risks from the excessive ingestion of Mo and Pb. Also, it would appear that the distribution of Co may be of value for delineating areas in which the risks of Co deficiency are high in animals. Similarly, it is reasonable to expect that geochemical information can contribute to the recognition of high risk areas for Se deficiency and for those in which excesses of Cd, Zn, As, F, Fe, and Ca may be involved in the aetiology of disease.

Geochemical maps cannot be used directly because of the influence of other variables; nevertheless they can be of value if factors modifying element pathways to animals are considered. Geochemical maps have the advantage of providing comprehensive spatial data including elements essential for animal nutrition, and those that are potentially toxic or interfere with the absorption or utilisation of essential elements.

In the United Kingdom the incidence of such diseases is not declining, indicating that circumstances enhancing the risks of their development are inadequately identified. Geochemical maps may suggest lines of investigation which provide a direct indication of disease, whereas studies confined to clinical or biochemical investigations on animals, may contribute little towards understanding the aetiology of such problems.

Geochemical data may also suggest new lines of enquiry into the nature of antagonists that inhibit the utilisation of essential trace elements. For example, the variables inhibiting the utilisation of Cu, Zn and possibly Se by animals have not yet been fully identified. Variables influencing the utilisation of Cr, Si, Ni, V and several other essential trace elements also require investigation. Evidence of unusually low or high concentrations of potential antagonists in soils and their parent materials could influence investigations into the significance of elemental deficiencies or excesses upon animal health.

Few diseases in man can be linked with element deficiencies or toxicities. It is generally accepted, however, that there is a relationship between endemic goitre and iodine deficiency; also between fluorine deficiency and dental caries. But these examples cannot be fully explained in terms of simple deficiency. The association between geochemical variables and human health in Britain remains largely uninvestigated.

There is strong evidence of a water factor associated with cardiovascular mortality in Britain and other countries. The factor is correlated with water hardness and current research seeks to determine its nature and its mode of action.

In certain studies the Ca and Mg content of water may make a critical contribution to total intake of these major elements. Together these elements have been considered as the water factor in cardiovascular disease. The intake of Na is currently under suspicion as possibly being involved in increased blood pressure in young people.

The role of geochemistry in the aetiology of cancers is uncertain and very few correlative studies have been carried out.

The possibility of heavy metals derived from soils being involved in certain chronic degenerative diseases of the central nervous system has been raised and several metals (Cu, Pb, Mo) seem worthy of investigation.

Geochemical factors can increase the exposure of people to environmental lead - acid water can dissolve lead from plumbing, and children in particular may ingest lead from soil and dust. Lead may be involved in cardiovascular disease; evidence of intellectual impairment in children due to low level lead exposure is inconclusive. There are a number of trace metals (Cd, Zn, Cr, and Se) for which there is some evidence that they may be involved in disease processes. However, very little research has been carried out in this field.

Studies relating geochemistry to human health will depend to a considerable degree on measurement of the amount of various trace elements in blood. But this is complicated both by sampling and analytical error and improved standards are essential if future studies are to be succesful.

7. RECOMMENDATIONS

The current geochemical mapping programme of the Institute of Geological Sciences should continue with the aim of providing systematic point source data for a wide range of chemical elements over the whole of Britain.

Research into the application of data processing techniques should be encouraged to develop interactive statistical programmes for the comparison of geochemical data with agricultural and epidemiological data as a first step in (a) recognising relationships between levels of chemical elements and disease incidence; (b) constructing predictive models of likely disease incidence based on geochemical data and (c) assessing the degrees of accuracy required of such data.

Research into the interpretation of geochemical reconnaissance maps should continue to emphasise the importance of chemical and physical processes influencing the distribution of elements in the rock-stream sediment-soil system, and in particular into the chemical forms of elements in the surface environment.

Systematic information on the distribution and forms of trace elements in soils is required as a prerequisite for predictive studies into plant, animal and human health.

It is important to obtain a better understanding of the physical chemistry of trace metals in soils, and in particular the sorption-desorption relationships for the major soil types in Britain.

A clearer understanding of complexing ligands in natural and sludge-amended soils is important. Also, it is relevant to know whether changes in agronomic practice can greatly change the production or proportions of these compounds and thus alter trace metal availability to plants.

Information is required on the influence of biomass turnover in the soil on the availability of trace elements.

Investigations are necessary to establish rates of trace element removal by modern high-yielding crops.

The relationship between genetic composition and the concentration of elements in the plant shoot should be studied as this can affect the plants' nutritional acceptability for livestock or human consumption. Major new cultivars should be thoroughly tested for metal uptake when grown in normal and high metal soils. Mechanisms at the root-soil interface which control trace element uptake also need more attention as they may be concerned with tolerance mechanisms.

Further case histories linking tne geochemical environment and animal health/disease are essential in order to assess the full potential of geochemical surveys to predict areas in which clinical and subclinical manifestations of disease due to trace element deficiency or excess are significant. Future studies should be statistically sound and take into account the possible multicausal nature of some problems.

In particular it is important to examine quantitatively the relationship between geochemical data for Ms and Cu and the response of animals. Studies should take into account the influence of major variations in soil conditions, such as pH and drainage, upon such relationships.

In view of the economic importance of Se deficiency to livestock, investigations into the geochemical distribution of this element are a priority. Information on the geographical distribution of Se and the relationship to low contents of Se in soils and crops is now required. Areas within which the risks of development of Se deficiency are high are poorly defined. Thus relationships between the geochemical distribution of Se, crop Se content and the incidence of Se responsive diseases should be investigated. Research should take into account soil variables such as sulphate and Fe content that may affect the availability of Se to plants; the significance of these variables as determinants of crop Se content under practical farming conditions is largely unknown.

Investigation of relationships between the geochemical distribution of F, I, and Na and that of fluorosis, I deficiency and Na deficiency in livestock should be encouraged. Although losses attributable to these diseases within Britain are probably small, they are severe in certain other countries. The potential of geochemical surveys for identifying high risk areas overseas should be investigated with financial and technical support from the United Kingdom.

Geochemical data are required for P distribution in areas with acid soils developed from granite rock to assist in the more effective delineation of areas in which P deficiency may be found in sheep.

The United Kingdom has, in most respects, an ideal geochemical background within which to undertake such studies. It also has almost unique expertise for the investigation of the effect of chemical anomalies upon the composition of crops used as foods or upon the composition of animal tissues and for studying relevant pathological changes in animals. It is recommended that this should be used, not only in support of investigations relevant to problems of the animal industry of the United Kingdom, but also to determine whether geochemical studies could contribute towards resolution of the more extensive and severe problems encountered within many developing countries. Investigations based initially upon a geochemical survey with its inherent advantage of rapid area coverage could be of particular value in land use planning as well as to the livestock industries.

The role of geochemical factors in the aetiology of human disease needs further careful consideration. In particular it is necessary to establish (a) the causal factors linking cardiovascular disease mortality and the quality of drinking water, and (b) to what extent geochemical factors affect Pb intake by children and their intellectual development in the United Kingdom.

It is recommended that a standing multidisciplinary committee should be set up by the Royal Society which would meet on an ad hoc basis to consider the current state of (a) specific diseases in which geochemical influences are considered possible; (b) specific elements and their possible relationship to physiological mechanisms and specific disease. The committee would invite an individual or a group with appropriate expertise and experience to prepare a brief report on the situation regarding a particular disease or a particular element. The committee would review the prepared material and make recommendations regarding appropriate reserach. The committee would also organise discussion meetings on some issues to encourage debate among interested individuals and groups of scientists (geochemistry, epidemiology, biochemistry, nutrition, water, soil and plant scientists, environmental pollution).

Participation in international quality control programmes for human tissue analysis should be encouraged.

INDEX

Animal health
    and geochemistry 59-95
Aphosphorosis
    (see also phophorous deficiency) 89
Arsenic 7, 9, 29, 30, 68, 74
    essential to animals 64
    in soils 87
    in stream sediments 87
    interaction with iodine 76,87
    toxic effects in animals 66, 122

Basalt
    abundance of elements in 9-10
Bedrock geochemistry
    northern Scotland 27
    England and Wales 27-28
Black shales 11, 30
    of Namurian age 79
Blood
    trace elements in 117-118
Boron 47
    deficiency in crops 51-52
British Regional Heart Study 106-107

Cadmium 7, 9, 29, 30, 70, 74, 75, 77
    and cardiovascular disease 115
    and human health 114-115, 124
    derived from cigarette smoking 114-115
    effect of Ca on absorption 105
    effect on Cu metabolism 86
    in herbage 86
    in soils, vegetables and dusts at Halkyn 110-111
    in soils, vegetables and dusts at Shipham 110-111
    in stream sediments 86
    interacting components in diet 76
    toxic effects in animals 66, 70, 86, 123
Calcium 7, 8, 37-39, 67, 68, 70, 72, 73, 75, 88, 123
    and water hardness in Britain 104
    effect on absorption of Cd in water 105
    effect on absorption of Cr in water 105
    effect of absorption of Pb in water 105
    effect of absorption of Zn in water 105
    in drinking water 105-106

interaction with Mn 76
interaction with Pb 76
interaction with Zn 76
in water and cardiovascular disease 103-104
in water and human health 105-106
Calcium carbonate
adverse effect on animal growth 74, 88
inducing deficiencies of P, Fe and Mn 88
interaction with Co 76
Cancer
and geochemistry 108, 124
related to Zn: Cu ratio in soil 108
Cardiovascular disease
(see coronary heart disease, hypertension and stroke) 102-104, 129
and Ca in water 105-106, 124
and Cr in drinking water in Japan 103
and Mg in drinking water 106-107, 124
and Pb in water 113
and smoking 103-104
and water hardness 103-104, 124
and Zn in water 115
effects of Cd on 115
regional differences 102-104
Chromium 7, 9
and cancer 116
and cardiovascular disease 116
and human health 116-124
deficiency in animals 62
effect of Ca on absorption 105
interacting components in diet 122
Cigarette smoking
as a source of Cd 114-115
Cobalt 7, 9, 47, 52, 67, 68, 70, 73
coprecipitation with Mn 82
deficiency in animals 63
deficiency in sheep (also see pining) 81-82, 122
in soils 82
in stream sediments 81-82
interacting components of diet 76
Colloids 17-18
Copper 7, 9, 29, 30, 47, 52, 67, 68, 70, 73, 74, 75, 77, 124
deficiency in animals 62-63, 78-81
induced by Mo 79-81, 92-94, 95
deficiency in crops 54
effect of soil ingestion on 88-89
in soils 79
interacting components in diet 76, 122
interaction with dietary Fe 88-89
Coronary heart disease (see cardiovascular disease, hypertension and stroke) 102-104
Crop problems in Britain 51-54
related to geochemical surveys 122

Data processing 23, 93, 127
Deficiency diseases
in animals 62-65
with geochemical involvement 63-64
adaptation to 70-71
Dental caries
related to F 101-102, 123
related to Pb in soil and food 111-112
Derbyshire
relationship between Pb in soil and blood and hair Pb 112
Dietary components
effects of mining and smelting 110-112
inorganic composition 67-69
Dietary requirement
of animals 67, 68
Drinking water
and cardiovascular disease 103-104, 129
as supplier of Ca 105
as supplier of Cd 114
as supplier of Cr 116
as supplier of Mg 106-107
as supplier of Na 108
as supplier of Pb 113
as supplier of Se 117
as supplier of Zn 115
at Shipham and Halkyn 110

Earth's crust
abundance of elements in 9-10
Eh 16-17, 26, 50
Eh/pH diagrams 16-17
Environmental geochemistry
definition 7
distribution of elements in rocks 7-13
principles 5-32

Fluorine 7, 10, 70
and dental caries 101-102, 123
and tooth mottling 102
as a carcinogen 102
essential to animals 64
human health 101-102
in drinking water 87
interacting components in diet 76
in water related to cancer 102
toxic effects in animals 66, 87, 123, 129
Fluorosis
(see also fluorine) 87, 129
mottling of teeth 102
skeletal 102

Food
    as supplier of Cd 114
    as supplier of Cr 116
    as supplier of Se 117
    as supplier of Zn 115
Fruit and vegetable consumption 111

Geochemical environment
    and cardiovascular disease 102-104
    response of animals to 66-76
Geochemical surveys
    by Imperial College, London 20-23, 24, 121
Geochemical surveys
    by Institute of Geological Sciences 20-23, 24, 121
Geochemical surveys
    prediction of animal deficiency and toxicity 77-89, 122, 123, 128
        arsenic toxicity 87-88, 122
        cadmium/zinc toxicity 86, 122
        calcium excess 123
        cobalt deficiency 81-83, 122
        copper deficiency 78-81, 92-94, 95
        fluorine toxicity 87, 122
        iron excess 123
        lead toxicity 85, 122
        molybdenum excess 122
        selenium deficiency 83-84, 122, 128
        selenium toxicity 84
        zinc and manganese deficiencies 84-85
Geochemical surveys
    Argyll 24
Geochemical surveys
    England and Wales 22, 24
Geochemical surveys
    Great Glen 24
Geochemical surveys
    Lewis - Little Minch 24
Geochemical surveys
    Moray - Buchan 24
Geochemical surveys
    Northern Ireland 22, 24
Geochemical surveys
    Orkney 24
Geochemical surveys
    Shetland 24
Geochemical surveys
    South Orkney and Caithness 24, 92-94
Geochemical surveys
    Sutherland 24
Geochemistry 7
    related to animal health 59-95
    related to cancer 108

related to cardiovascular diseases 102-104
related to multiple sclerosis 109
soil information relating to 55-56
Goitre 101-123
Granite
abundance of elements in 9-10
associated with Co deficiency in sheep 81
associated with P deficiency in sheep 129
associated with Se deficiency in sheep 83
Granodiorite
abundance of elements in 9-10

Halkyn
Cd and Pb in soils, vegetables and dust 110-111
health studies in 110-111
Hard water 26
and cardiovascular disease 103-104
chemical properties of 15
Human health
and geochemistry 97-119
Human tissue analysis 130
Hypertension
(see cardiovascular disease, coronary heart disease and stroke)
102-104
Cd in kidneys 115

Igneous rocks
abundance of elements in 8, 9-10
Imbalance of inorganic elements
influence on availability and tolerance 71
Intelligence
effects of Pb on 113-114, 124, 129
Interactions between inorganic elements
influence on function 74
influence on tissue distribution 74
influence on toxicity 74
Iodine 7, 10, 67, 73
deficiency in animals 63, 129
human health 101, 123
interacting components in diet 76
Iron 7, 47, 50, 52, 67, 68, 70, 72, 73, 75, 77
effect on Se availability 129
interacting components in diet 77
toxic effects in animals 66, 123

Lead 7, 10, 29, 30, 68, 70, 75, 77
and human health 113-114, 124
effect of Ca on absorption 105
essential to animals 64
excessive ingestion by animals 122
in bovine blood 85

in herbage 85
in soils and dental caries 102
in soils, vegetables and dusts at Halkyn 110-111
in soils, vegetables and dusts at Shipham 110-111
in stream sediments 85
interacting components in diet 76
related to cardiovascular disease 113
related to children's intelligence 113-114, 124, 129
related to dental caries 111-112
relationship between soil, blood and hair Pb in Derbyshire 112
toxic effects in animals 68, 85
Limestone
abundance of elements in 9-10

Magnesium 7, 8, 37-39, 47, 67, 68, 69, 70, 73
interacting components in diet 76
in water and cardiovascular disease 103-104
supplies in drinking water 106-107
Major elements 7
essential to plants 37-39, 121
Manganese 7, 10, 47, 50, 52, 67, 68, 70, 72
deficiency in animals 63, 84
deficiency in crops 52-53
in herbage and cereal straw 84
interacting components in diet 76
Mercury 7, 10, 70
Metals
excess in food crops 30
pollution 32
Metal toxicity
in crops 30
in livestock 30
Metamorphic rocks
abundance of elements in 12-13
Mineralization 30
Mining and smelting
and human health 109-112
Molybdenum 7, 10, 29, 30, 47, 52, 68, 72, 74, 75, 77
and dental caries 101
and human health 109, 124
deficiency in animals 63
distribution in Caithness 92-94, 95
distribution in England and Wales 21, 52
excess ingestion by animals 122
induced copper deficiency in animals 66, 79-81, 92-93, 95, 128
in stream sediments 79-81, 92-94
interacting components in diet 76
stream sediment anomalies 79-81, 92-94
sulphide complexes 74
toxic effects in animals 66
Molybdenosis 79

Multiple sclerosis
    and geochemistry 109
    related to soil Cu, Zn, Pb and Mo 109, 124
    resemblance to swayback 109

Nickel 7, 10, 30, 68
    deficiency in livestock 89
    interacting components in diet 123
    relationship with N metabolism 89
Nitrogen 37-39, 69

Oxidation - reduction
    effect on trace element availability 48

pH 16-17, 26, 48, 128
Phosphorous 7, 37-39, 47, 67, 68, 73
    deficiency in animals 64, 89, 129
    effect of high intake on Fe, Mn, Ca and Zn 72
    effect of soil Fe ingestion on 88
    interacting components in diet 76
Pining
    (see also cobalt deficiency in sheep) 81-82
Plant-soil processes 35-56
Plumbosolvency 113
Potassium 7, 8, 37-39, 47, 68, 69, 73
    interaction with Mg 76

Radon 19
Rainwater
    chemical composition 14
Regional geochemical maps
    (see also trace element maps) 19-25, 121-122, 127
        and animal health 122-123
        and human health 109-112
Regional geochemistry of Britain 25-28
Rhyolite
    associated with Co deficiency in sheep 81
River water
    chemical composition 14
    Zn in United States 115
Royal Society Working Party on Environmental Geochemistry and Health
    actions taken 2
    conclusions 121-125
    membership 3-4
    origins and remit 1
    recommendations 127-130
Rural populations
    trace element intake 111

Sandstone
    abundance of elements in 9-10
    associated with Co deficiency in animals 81

Sedimentary rocks
abundance of elements in 8-12
chemistry 12
mineralogy 12
Selenium 7, 10, 30, 50, 52, 67, 68, 70, 74, 77
and breast cancer 117
and dental caries 101
and gastro-intestinal cancer 117
and human health 116-117, 124
and Keshan Disease 117
deficiency in animals 62, 64, 83-84, 122, 128
interacting components in diet 76, 122, 129
toxicity in animals (also see selenosis) 30, 66, 84
Selenosis 30, 66, 84
Sequestration
of elements in tissues 69-70
Shale 30
abundance of elements in 9-10
Shipham
Cd and Pb in soils, vegetables and dust 110-111
health studies in 110-111
Silicon 8, 68
deficiency in animals 64
interacting components in diet 123
Silver 9
interaction with Cu 76
interaction with Se 76
Sodium 7, 8, 67, 68, 73
deficiency in animals 64, 129
in drinking water 107
related to blood pressure 107
Soft water 25-26
and cardiovascular disease 103-104
chemical properties of 15
Soil
abundance of elements in 9-10
factors influencing availability 47-51
major elements in 37-39
rock-soil system 51
trace elements in 9-10, 28-32, 37
Soil ingestion 88
as supplier of As 87
as supplier of Cd and Zn 86
as supplier of F 87
as supplier of Fe 87
as supplier of Pb 85
influence on Cu nutrition 88-89
influence on P nutrition 88
Solubility
of elements in digestive tract 72-74
Solutions 15

Stroke
(see cardiovascular disease, coronary heart disease and hypertension) 102-104
Sulphur 67, 68, 72, 73
interaction with Pb 76
interaction with Se 76, 129
Surface environment
chemical elements in 15-19

Tamar Valley
health studies in 110-111
Toxicity of inorganic elements
in animals 65-66
adaptation to 70-71
Trace element maps
(see also regional geochemical maps) 19-25
and animal health 122-123
and human health 109-112
Trace elements 7
abundance 9-10
and cardiovascular disease 103-104
and human health 112-117
availability to plants 40-51, 121, 127
availability in the soil 47-50, 127
quantity and chemical form, effect of 47-49
root-soil system 51
valency, effect of 50
complexing with organic liquids 49, 127
composition of plant parts 45
crop offtake 46-47
essential to plants 37, 39-40, 121
factors influencing deficiencies in plants 52
in blood 117-118
influence of biomass turnover 127
influence of plant factors on uptake processes 40-41
element interactions 42
growth rate 43
root factors 43
species and genotype effects 43-45, 128
speciation 48
trace metal function in plants 46
Tungsten 10
interaction with Mo 76

Ultra basic rocks 8, 30
abundance of elements in 9-10
Uranium 10, 18-19

Vanadium 7, 10
essential to animals 64
in hydroxyapatite 102
interacting components in diet 123

Vitamin B_{12} 70

Wales
    health studies in 110-112
Weathering
    effect on redistribution of elements 13-14
    influence of bedrock 13
Wolfson Geochemical Atlas of England and Wales 21, 22

Zinc 10, 29, 30, 47, 52, 68, 69, 70, 72, 73, 74
    and human health 115-116, 124
    deficiency in animals 64, 84
    deficiency in crops 53
    effect of Ca on absorption 105
    effect on Cu metabolism 86
    in herbage and cereal straw 84
    in soil related to gastric cancer 108
    in stream sediments 86
    interacting components in diet 76, 122
    related to cardiovascular disease 115
    toxic effects in animals 66, 123
Zinc: copper ratio
    in soil related to gastric cancer 108